The increasing popularity of wireless networks makes interference and coexistence between multiple systems inevitable. This book describes techniques for quantifying this, and the effects on the performance of wireless networks operating in the same wave band. It also presents a variety of system-level solutions, showing the need for new hardware implementations. The book starts with basic concepts and wireless protocols before moving on to interference performance evaluation, interference modeling, coexistence solutions, and concluding with common misconceptions and pitfalls. The theory is illustrated by reference to real-world systems such as Bluetooth and WiFi. With a number of case studies and many illustrations, this book will be of interest to graduate students in electrical engineering and computer science, to practitioners designing new WLAN and WPAN systems or developing new techniques for interference suppression, and to general users of emerging wireless technologies.

Nada Golmie received her Ph.D. in Computer Science from the University of Maryland, College Park, in 2002. Since 1993 she has been a research engineer for the advanced networking technologies division at the National Institute of Standards and Technology (NIST). This publication interest is not affiliated with her work there. Nada's research has led to over 100 papers published in professional conference proceedings and journals, and has contributed to industry IEEE standards.

Coexistence in Wireless Networks

Challenges and System-Level Solutions in the Unlicensed Bands

Nada Golmie

CAMBRIDGE
UNIVERSITY PRESS

CAMBRIDGE UNIVERSITY PRESS
Cambridge, New York, Melbourne, Madrid, Cape Town, Singapore, São Paulo

Cambridge University Press
The Edinburgh Building, Cambridge CB2 2RU, UK

Published in the United States of America by Cambridge University Press, New York

www.cambridge.org
Information on this title: www.cambridge.org/9780521857680

First published 2006

Printed in the United Kingdom at the University Press, Cambridge

A catalog record for this publication is available from the British Library

ISBN-13 978-0-521-85768-0 hardback
ISBN-10 0-521-85768-6 hardback

To my parents
To Geof

Contents

Illustrations

Tables

Preface

Wireless networks are rapidly becoming a part of the ubiquitous computing environment, and whether they are enterprise networks or in public hot spots (for example in airports, hotels, homes), often they are deployed in infrastructureless environments. The rapid specification development phase and the tight time to market cycle that follows leave little room for performance enhancements and proper coexistence consideration.

Why did I write this book?

Having gone through a somewhat complete performance analysis and coexistence development cycle for wireless network technologies being developed by the IEEE 802 standard working groups, and having gained some experience on the topic, I feel compelled to share it with other network engineers and researchers that are pursuing similar objectives. In particular, I would like to share the methodologies developed and the lessons learned from this process with others embarking on a similar quest.

The audience for this book includes: (1) researchers interested in performance evaluation and interference mitigation techniques; (2) wireless systems engineers and practitioners designing wireless communication systems; (3) users of wireless networks.

This book is unique because it focuses on a system level view of the problem of interference and its solution space. Generally, interference is dealt with at the physical layer. There are several outstanding books that focus on the accurate characterization of the wireless channel in addition to the development of physical layer techniques for filtering and anti-jamming. By considering a system level view for the interference problem and providing methods for evaluating interference at the system level including the medium access control layer, the transport, and the application layers, this effort is intended to (1) complement previous work aimed at the physical layer, (2) provide more accurate tools for quantifying interference and its impact on the system performance, and (3) develop a catalog of coexistence techniques that could be applied in various environments and for different technologies.

About the organization of this book

The structure of this book is as follows.

Chapter 1 provides and introduction to the subject of coexistence in wireless networks.

Chapter 2 covers some of the basic concepts and the design choices in the development of physical and media access control layer protocols. Examples for select technologies of interest, such as Bluetooth and IEEE 802.11b, are given in order to illustrate the concepts described.

Chapter 3 defines an interference model in the context of this book. Some of the major building blocks for conducting a performance analysis study are discussed, including the terminology, performance metrics, and the system parameters affecting the performance.

Chapter 4 consists of an open-loop interference model. In this chapter approximations for evaluating interference at the victim's receiver while ignoring the interactions between the interferer and the victim system are given. This open-loop evaluation consists of a mathematical analysis technique based on a probability of packet collision in time and frequency.

Chapter 5 describes a closed-loop modeling environment in order to capture the mutual effects of interference on each end device including the protocol interactions. A modeling and simulation methodology is presented that includes the number of devices and their placement on a two-dimensional grid, setting the transmission power of the radios, and defining the traffic distribution pertaining to the applications considered.

Chapter 6 discusses channel estimation and selection techniques that constitute the basis for most interference mitigation techniques. Several metrics for channel estimation are presented and explained, including packet or frame loss, received signal strength, and packet acknowledgment. These metrics are then used as criteria for channel selection and interference avoidance.

Chapter 7 describes proven techniques to mitigate interference. The emphasis is on dynamic and system level mechanisms that are able to adapt to the interference environment. The techniques overviewed can be broken up into two broad categories. The first category of solutions consists of temporal and/or spectral sharing of the spectrum. The second category of solutions is about adaptation and the possibility of choosing either the radio or the network that is best suited to the environment.

Chapter 8 includes some common myths and misconceptions associated with interference mitigation solutions. The main goal is to shed some light on the lessons learned while researching and developing solutions.

Acknowledgments

I would like to thank David Su at the National Institute of Standards and Technology, who has supported my professional endeavors. I would also like to thank my colleagues at NIST and many exceptional people whom I collaborated with over the years. Their enthusiasm and dedication provided an enriching and stimulating environment for me to learn from and pursue research.

I am indebted to my dad for always believing in me. His pride, love, and encouragement have been and continue to be a constant source of inspiration in my life.

I would like to thank two people who hold a special place in my life. I am especially grateful to my mom, Raja, and my husband, Geof, for their unconditional love, infinite support, and for putting up with me during the writing of this book. It is due to their constant encouragement that I have been able to complete this project.

My gratitude goes to Phil Meyler from Cambridge University Press, whose idea it was for me to write this book and who made it all happen with his very able team of technical and production editors, including Emily Yossarian, Anna Littlewood, Emma Pearce and Irene Pizzie.

Abbreviations

ACK	acknowledgment message
ACKLoss	acknowledgment loss
ACL	asynchronous connectionless link also in Bluetooth
AFH	adaptive frequency hopping
AP	access point
API	application program interface
ARQ	automatic repeat request
AWGN	additive white Gaussian noise
BER	bit error rate
BIAS	Bluetooth interference aware scheduling
BT	Bluetooth technology
CCA	clear channel assessment (used in 802.11)
CCK	complementary code keying
CDMA	code division multiple access
CPU	central processing unit
CRC	cyclic redundancy check
CSMA/CA	carrier sense multiple access/collision avoidance
CTS	clear-to-send
DCF	distributed coordination function (in 802.11)
DIFS	DCF interframe space
DS	direct sequence spread spectrum
FCS	frame check sequence
FDM	frequency division multiplexing
FDMA	frequency division multiple access
FEC	forward error correction
FER	frame error rate
FH	frequency hopping spread spectrum
FIFO	first in first out queueing
FST	frequency status table
GFSK	Gaussian frequency shift keying
GHz	gigahertz

GMSK	Gaussian minimum shift keying
GPS	global positioning system
GSM	global system for mobile communications
IEEE 802.11	a set of IEEE standards for WLANs
IEEE 802.15	a set of IEEE standards for WPANs
ISM	industrial, scientific, and medical frequency band
LD	limiter–discriminator receiver
LIFO	last in first out queueing
LLC	logical link control layer
LMP	link management protocol
MAC	medium access control layer
MANET	mobile ad hoc network
MHz	megahertz
NACK	negative ACK
OFDM	orthogonal frequency division multiplexing
PDF	probability density function
PDU	payload data unit
PHY	physical layer
PLoss	packet loss
QOS	quality of service
RF	radio frequency
RR	round robin scheduling
RSNI	received signal to noise indicator
RSSI	received signal strength indicator
RTS	request-to-send
SCO	synchronous connection-oriented link in Bluetooth
SIFS	short interframe space
SIR	signal to interference ratio
SNR	signal to noise ratio
TCP/IP	transmission control protocol/internet protocol
TDM	time division multiplexing
TDMA	time division multiple access
TPL	transmit power level
UMTS	universal mobile telecommunications system
UNII	unlicensed national information infrastructure band
Wi-Fi	wireless fidelity, generic designation for an IEEE 802.11 network
WLAN	wireless local area network
WMAN	wireless metropolitan area network
WPAN	wireless personal area network

1 Introduction

The main themes of this book are to explore evaluation methods for quantifying the mutual effects of interference on the performance of wireless networks and to investigate system-level solutions for their coexistence in the same environment.

The coexistence of wireless communication systems operating in the same environment has become a "hot" topic in recent years as more systems are choosing to use the unlicensed bands and forfeiting the need to purchase spectrum.

There are two specified unlicensed bands for the operation of wireless systems, namely:

(i) the industrial scientific and medical (ISM) band that includes the 900 MHz, 2.4 GHz, and 5.8 GHz frequencies;

(ii) the unlicensed national information infrastructure (UNII) band that includes the 5.2 GHz band. This band was opened in 1997 in the United States in order to expand broadband access opportunities.

Few rules apply in the unlicensed bands such as the ISM band. For example, the rules defined in the Federal Communications Commission Title 47 of the Code for Federal Regulations Part 15 [2] relate to the total radiated power and the use of the spread spectrum and frequency hopping modulations. It is commonly understood that all users of the unlicensed bands can equally affect the quality and the usefulness of this spectrum. Thus, the major downside of the unlicensed band is that frequencies must be shared and potential interference tolerated.

We distinguish between several types of users in these unlicensed bands. Apart from emerging wireless networks, users include low cost devices such as video/audio transmitters for entertainment, security and surveillance, microwave ovens, and broadcast links for high power FM television.

Although the discussion and the examples provided in this book relate to wireless networks, the performance evaluation approach and the solutions may apply to other wireless systems.

There are three types of wireless networks that we consider depending on the bandwidth and the coverage area supported. Wireless personal area networks (WPANs) are intended for cable replacement systems and short distance ad hoc connectivity. Communications in WPAN are normally confined to a person or object and extend up to 10 meters in all directions. WPAN specifications include infrared [15], Bluetooth [1,7], Zigbee [11], and IEEE 802.15.3 [10]. This is in contrast to wireless local area networks (WLANs) that typically cover a moderately sized geographic area such as a single building or campus. WLANs operate in the 100 meter range and are intended to augment rather than replace traditional wired LANs. They are often used to provide the final few feet of connectivity between the main network and the user. WLAN specifications include Home RF [3] and the family of IEEE 802.11a/b/g standards [5,6,8]. Finally, wireless metropolitan area networks (WMANs) [14] are mainly designed for broadband connections over long distances (up to several tens of kilometers). Although they can be used to provide last mile connectivity to mobile and vehicular users, they are mainly intended for interconnecting WLAN hotspots and cellular coverage areas.

Thus, each wireless network type may be seen as filling a specific niche area and supporting a different application need, altough the co-location and simultaneous operation of all these networks in the same environment poses an unprecedented challenge since they are all competing for the same spectrum.

During the decade 1995–2005 we witnessed the emergence of a few dominant wireless technologies, such as IEEE 802.11b and Bluetooth, with more or less distinct requirements; the future for wireless networks will most likely combine different technologies in order to support constantly changing and evolving usage models and applications. For example, WPAN can be used to connect a headset or PDA to a desktop computer, which in turn may be using WLAN to connect to an access point placed several meters away that is connected to a WMAN deployed in the city.

The vision for interconnecting heterogeneous networks makes the coexistence problem extremely important and the solutions considered even more challenging. If the availability of the unlicensed bands makes the proliferation of wireless networks at all possible, coexistence is the only strategy to ensure their proper operation.

1.1   Interference modeling and performance evaluation

Since one of the objectives for this book is to identify methodologies for quantifying the effects of interference on network performance, a few observations are in order regarding the state of the art in the assessment of interference.

Published results can be generally classified into at least three categories depending on whether they rely on analysis, simulation, or experimental measurements in order to provide quantitative measurements.

1.1.1   Mathematical modeling

Analytical results are mainly based on modeling the collision of packets from multiple transmitters at the receiver and computing a corresponding packet error probability

[18,23,25,26,44,51,71]. This error probability is a function of several parameters; for example, the number of transmitters, the distance between the transmitters and the receiver, the difference in power level between the transmitter and the interferers, and the receiver technology considered. The results obtained are generally useful in order to gain a first order approximation on the impact of interference and the resulting performance degradation. However, these analytical models often make assumptions concerning the traffic distributions and the operation of the media access protocol which can make them less realistic. More importantly, in order for the analysis to be tractable, mutual interference that can change the traffic distribution for each system is often ignored. Therefore, mathematical modeling is often used to complement measurements obtained from experimental and simulation data.

1.1.2 Experimental modeling

Results obtained from experimental modeling are considered by far the most accurate at the cost of being too specific to the implementation tested. Examples of experimental measurements can be found in refs[21,34,40,52,61]. In the case where the implementation details are completely known, including various optional add-ons and parameters, then the evaluation can be extremely informative. However, access to the complete implementation details is often restricted by the vendors or the implementors leading to the so-called testing of black box equipment. Thus, the results obtained are not applicable outside the experimental set-up of the equipment tested. Furthermore, since parameters cannot be modified, their effects on performance is not easily understood. Therefore, experimental modeling is useful mostly in the context of specific product development and testing.

1.1.3 Simulation modeling

Simulation modeling constitutes a third alternative to the other two approaches described above. It consists of using computer simulations to model the behavior of the protocols under consideration. This approach can provide a flexible framework where detailed parametrized models for the media access control and physical layer protocols are combined and the interactions between the various system parameters are identified and accurately quantified. Simulation modeling is ideal for evaluating numerous "what if" scenarios without the cost associated with building and testing the equipment. Simulations play a critical role in evaluating scalability issues and complex system behavior where parameters are modified and their effects on the overall performance quantified. Examples for simulation models developed to evaluate wireless network interference can be found in refs [20,48,55,67].

The question we ask here is, how do these methods relate to the performance analysis techniques discussed in this book? Basically, the evaluation approaches we consider are broken up into two categories. First, we consider an open-loop interference evaluation technique where the effects of mutual interference are ignored. Secondly, we describe a closed-loop evaluation approach where the interactions amongst interfering systems are considered. Results comparing both approaches are also discussed. Observe that all three modeling approaches presented above can be used in either open-loop or closed-loop evaluations, although closed-loop modeling is often associated with simulation modeling and open-loop evaluation is more related to mathematical modeling.

1.2 Interference avoidance and coexistence strategies

As far as meeting the second objective for this book and exploring coexistence strategies, the second half of the book is devoted entirely to exploring the solution space. The focus is on adaptive and system-level solutions that can augment or enhance traditional filtering, anti-jamming, and physical layer techniques. Basically the emphasis is placed on solutions that do not require major changes to the hardware and the technical specifications of the technologies considered.

Interference mitigation has always been and remains a big part of any communication system design cycle. Since wireless system engineers have always had to contend with interference from both natural sources and other users of the medium, the classical communication design cycle has consisted of predicting channel impairments and choosing adequate modulation and error correction schemes. Error correction can even be made to be adaptive to the error characteristics in the operation environment, as was shown by Eckhardt and Steenkiste [35], in which the effects of using an adaptive error correction scheme are investigated and adaptive schemes adjusted based on the environment. Power control is another adaptive technique generally used in spread spectrum systems such as carrier division multiple access (CDMA). In addition to these design choices and adaptive techniques, there are several known physical layer interference suppression techniques such as notch filtering and adaptive equalization [56]. In contrast to these so-called classical approaches to interference mitigaton, our contribution becomes valuable when redesigning systems from scratch is not considered to be a viable option. Therefore, we favor in our discussion adaptive control strategies, system parameter adjustments over other signal processing, and physical layer strategies that are well documented and widely available in the literature. The techniques presented in this book are meant to complement what is generally done in signal processing at the physical layer.

1.2.1 Industry led activities

There are few industry led activities tackling the issue of coexistence. Two efforts that we mention here are under the auspices of the IEEE 802 LAN/MAN standards committee that develops local and metropolitan area network specifications. The IEEE 802.19 Technical Advisory Group (TAG) on coexistence was formed in 2003 in order to develop and maintain policies defining the responsibilities of IEEE 802 standards working groups regarding coexistence. This is a standing group that advises the 802 executive committee on coexistence matters and assists various 802 working groups to assess and develop coexistence strategies accurately. Prior to the formation of the IEEE 802.19 TAG, the IEEE 802.15.2 Task Group on coexistence published a recommended practices document [9] for the coexistence of Bluetooth and IEEE 802.11b devices. This document considers solutions for mitigating the interference between these two technologies. Solutions range from collaborative schemes to be implemented in the same device to fully independent solutions that rely on interference detection and estimation.

- Mechanisms for collaborative schemes are based on a MAC time domain solution that alternates the transmission of Bluetooth and IEEE 802.11 packets (assuming both protocols are implemented in the same device and use a common transmitter) [22]. A priority of access is given to Bluetooth for transmitting voice packets, while WLAN is given priority for transmitting data.
- The non-collaborative mechanisms considered range from adaptive frequency hopping [24] to packet scheduling and traffic control [43]. They all use similar techniques for detecting the presence of other devices in the band, such as measuring the bit or frame error rate, the signal strength or the signal to interference ratio (often implemented as the received signal indicator strength (RSSI)). Other MAC scheduling techniques known as packet encapsulation rules [19], or overlap avoidance (OLA) [32], use the variety of Bluetooth packet lengths to avoid the overlap in frequency between 802.11 and Bluetooth. In other words, the Bluetooth scheduler knows to use the packet length of proper duration in order to skip the so-called "bad" frequency.

1.2.2 Fair scheduling and wireless QOS research

A research topic that has received more attention recently and is closely tied to wireless coexistence is fair scheduling and the support of quality of service (QOS) requirements in a wireless environment.

For example, Fragouli *et al.* [38] propose a strategy that combines class-based queuing [36] with channel-state based scheduling [31] that eliminates the head of line problem caused by first in first out (FIFO) queuing when certain devices suffer from a bad link. In ref. [38], link sharing guidelines are provided to maximize channel utilization and limit the access of misbehaving sources.

Furthermore, a number of algorithms have been proposed on fair scheduling [57,58,62]. While there may be some differences in implementation and complexity, the basic idea in all these algorithms is for sources experiencing a bad wireless link to relinquish the unutilized bandwidth to other sources that can take advantage of it. Compensation in bandwidth occurs when the channel conditions improve in order to achieve the so-called long term fairness objective.

While the interference mitigation problem that we are trying to solve bears some resemblance to some of the problems addressed in refs [38,57,58,62], there are important differences to note. Regarding interference mitigation, it is important to consider an instantaneous measure of fairness rather than a long term fairness objective. The reason is as follows. All previous work uses a two state Markov channel model for each link. The transition probabilities between the good and bad states are in the order of several seconds to account for periods of fading, multipath and various other wireless effects. The situation in the case of interference is somewhat different due to the interactions of the systems. Since different devices in the same piconet will be subject to different interference levels due to parameters such as geometry and transmitted power, not all frequencies will be equally good to all devices. Therefore, the goal is to assign frequencies optimally such as to maximize channel utilization and guarantee fairness among all the devices.

2 Basic concepts and wireless protocol overview

This chapter is designed to give the reader a comprehensive understanding of the fundamentals in wireless protocol design. First, we overview some of the physical layer and the medium access control layer design choices. Then, we give the details of select major protocols as examples of the concepts described.

2.1 Physical layer

The physical layer has the main function of transporting the information bits passed by the higher layers over a physical medium and recovering them on the other side of the medium. We can view the physical layer in terms of a digital or analog communication channel and modules that map digital information to an analog signal in case the channel is analog. Figure 2.1 illustrates the main components of the physical layer that are discussed in the following sections. For an in-depth treatment of communication systems, the reader is referred to other texts [70,75].

2.1.1 Communication channel

A communication channel consists of a physical medium, such as radio waves, copper wire, optical fiber, and the associated equipment necessary to transmit information over the medium. Communication channels can be used for either digital or analog transmission. Digital transmission consists of transmitting a sequence of pulses corresponding to a sequence of information bits. Analog transmission involves the transmission of waveforms associated with the transmitted signal. The bandwidth of a channel, W, measures the width of the window of frequencies that are passed by the channel. A low-pass communication channel passes low frequency components, while a bandpass channel passes power in some frequency range f_1 to f_2. The bandwidth of the channel, W, is thus equivalent to $f_2 - f_1$. In order to modify the frequency components of a signal, filters are commonly used in the transmitter and the receiver circuitry.

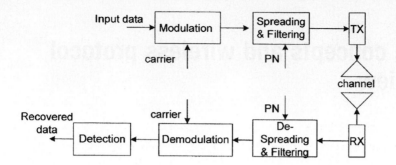

Fig. 2.1. Physical layer system components.

2.1.2 Modulation and filtering

Modulation is required to map digital information into a waveform sent over an analog channel. Digital modulation is the process of transforming a group of k bits, also called symbols, into waveforms. There are $2^k = M$ symbols in an alphabet. A waveform is expressed as follows:

$$s(t) = A(t)\cos[w_0 t + \phi(t)] \tag{2.1}$$

where $A(t)$ is the signal amplitude, $\phi(t)$ is the angle or phase, and w_0 is the center frequency.

There are three basic modulation techniques known as amplitude shift keying, frequency shift keying, and phase shift keying.

In amplitude shift keying the signal's amplitude is varied in order to encode M symbols:

$$s_i(t) = A_i(t)\cos[w_0 t + \phi] \quad \text{for} \quad i = 1, \dots, M \tag{2.2}$$

Similarly, for frequency shift keying, the frequency is varied according to

$$s_i(t) = A\cos[w_i(t) + \phi] \quad \text{for} \quad i = 1, \dots, M \tag{2.3}$$

And finally for phase shift keying, the phase or angle is varied:

$$s_i(t) = A\cos[w_0 t + \phi_i(t)] \quad \text{for} \quad i = 1, \dots, M \tag{2.4}$$

This process of mapping the information bits into waveforms and transmitting them over a low-pass communication channel is also known as baseband modulation. However, additional signal processing is required in order to match the signal to be transmitted with the channel characteristics, or the bandpass channel. Thus, a bandpass modulation is a baseband modulation whose spectrum has been shifted to a frequency band passed by the channel considered.

The basic function is to transmit a low-pass signal $I(t)$ over a bandpass channel centered at f_c. In order to translate the spectrum of $I(t)$ to another signal centered around f_c, we can multiply it with a carrier $C(t)$ of the following form:

$$C(t) = A \cos(2\pi f_c t + \theta) \tag{2.5}$$

where A is the amplitude of the carrier, f_c is the carrier frequency, and θ is an arbitrary phase constant. The product $X(t)$ has the following form:

$$X(t) = I(t)C(t) = AI(t) \cos(2\pi f_c t + \theta) \tag{2.6}$$

Filtering is performed throughout a communication system and for a variety of reasons. The primary reasons to adopt filtering is to select the desired signal, minimize the effects of noise and interference, modify the spectra of signals, and shape the time-domain properties of digital waveforms in order to improve their detectability. For example, receivers use filters to reject out-of-band noise, while transmitters use filters to meet regulatory constraints on the shape of the spectra transmitted. In some cases, filters are required to be adaptive in terms of changing their response to changing properties of the signal. A filter is introduced to remove signal distortions introduced by a channel as it is expected to change its reponse as the channel characteristics change. The adaptive tapped delay line is the most common adaptive filter structure, also referred to as an equalizer.

2.1.3   Channel propagation properties

The modulated signal propagates in a medium at a speed of v m/s, where

$$v = \frac{c}{\sqrt{ef_0}} \tag{2.7}$$

and where $c = 3 \times 10^8$ m/s is the speed of light in a vaccum and e is the dielectric constant of the medium. In free space, $e = 1$, and $e \geq 1$ otherwise. The wavelength λ of the signal is given by the length in space spanned by one period of the sinusoidal signal:

$$\lambda = \frac{v}{f_0} m \tag{2.8}$$

The modulated signal is also attenuated as it travels through the media. The attenuation in wireless media is proportional to d^n, where d is the distance travelled and n is the path loss exponent; $n = 2$ for free space and for environments where obstructions are present $n > 2$. The attenuation in decibels (dB) is proportional to $n \log_{10} d$ dB. The attenuation phenomenon is determined at the receiver in terms of the signal to noise ratio (SNR) and bit error rate (BER).

2.1.4   Signal detection

The transmitted signal is recovered at the receiver as follows. Any signal can be expressed in terms of a linear combination of N orthogonal waveforms:

$$s_i(t) = \sum_{j=1}^{N} a_{ij} \phi_j(t); \quad i = 1, \ldots, M; \quad N \leq M \tag{2.9}$$

Let s_j for $j = 1, \ldots, k$ represent reference signals belonging to a set of M waveforms. The received signal r is $r = s_j + n$, where n represents noise. The receiver has to decide whether r closely resembles either of the reference signals s_j by measuring the distance of r with all reference signals s_j as illustrated in Figure 2.2. The chosen reference signal s_j is the one that leads to the minimum distance, $d = ||r - s_j||$.

Additionally, in coherent detection, the receiver exploits the knowledge of the carrier's phase, while in non-coherent detection, no phase information is used. Further reading on coherent detection is found in refs [70,75].

2.1.5   Spread spectrum

Spread spectrum was originally developed in the 1940s for military communications in order to make the signal less obvious to enemy interception and jamming capabilities. The basic idea is to transmit the signal over additional bandwidth, using less power per frequency, but more frequencies. Thus the information signal $s(t)$ is multiplied by a pseudo-noise (PN) signal $c(t)$

$$m(t) = c(t)s(t) \tag{2.10}$$

and the resulting signal $m(t)$ has the same wideband characteristics as the PN signal, as illustrated in Figure 2.3. Thus each bit in the original digital signal $s(t)$ is chopped

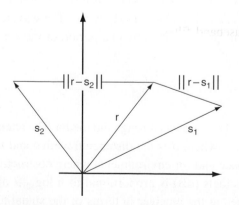

Fig. 2.2. Detection at receiver.

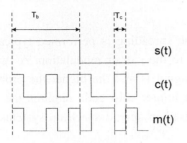

Fig. 2.3. Illustration of spread spectrum.

up into small time increments commonly referred to as "chips". If T_b is the duration of a bit in time and T_c is the chip duration, then the spread factor N is given by

$$N = \frac{T_b}{T_c} \tag{2.11}$$

The received signal $r(t)$ consists of the transmitted signal $m(t)$ and some additional interference signal $i(t)$:

$$r(t) = c(t)s(t) + i(t) \tag{2.12}$$

By multiplying the received signal by the same PN signal $c(t)$ we obtain

$$z(t) = c^2(t)s(t) + c(t)i(t) \tag{2.13}$$

Since the PN signal consists of alternating $+1$ and -1 levels, $c^2(t) = 1$ for all t. Therefore, the recovered signal $z(t)$ becomes

$$z(t) = s(t) + c(t)i(t) \tag{2.14}$$

As $z(t)$ is further processed by a baseband filter to retrieve the narrowband signal $s(t)$, the wideband interference component $c(t)i(t)$ can be filtered out. In essence, the same spreading effect is applied on the interfering signal at the receiver as it is applied on the information signal at the transmitter. This widening effect of the interference makes it easier to be filtered out.

Note also that in order to recover the signal, a receiver has to use the same PN signal that was used in the transmission. To a third party that does not have the spreading parameters, the signal appears like noise.

There are two main variations for spread spectrum that are commonly used in wireless network technologies. They include direct sequence and frequency hopping spread spectrum.

Direct sequence spread spectrum

In direct sequence (DS) spread spectrum, the spreading is performed typically, as described in the preceding paragraphs, prior to the carrier modulation. As we saw, the longer the chip code, the stealthier the signal is and the higher the probability of recovering it at the receiver. However, a longer chip code requires the use of additional bandwidth. A common measure used in spread spectrum systems is the processing gain G defined in dB as

$$G = 10 \log_{10}(r_c/r_b) \tag{2.15}$$

where r_c is the chip rate and r_b is the bit rate of a system. For example, if $r_c = 11$ Mchip/s and $r_b = 10$ Mbit/s, the processing gain $G = 10.41$ dB. Note that DS technologies are used in IEEE 802.11b [6] PHY layer and IEEE 802.15.4 [11] specifications.

Frequency hopping

While in the direct sequence spread spectrum, the spreading effect is achieved using a PN signal; in frequency hopping systems frequencies are randomly selected over time in order to cover a wide spectrum. This method makes it possible to cover a wider spectrum than that used in direct sequence. The frequency occupancy pattern over time forms the so-called frequency hopping sequence. Frequency hopping sequences are generally generated randomly. They can even be selected dynamically in order to avoid interfering with other signals in the same band. An important measure to consider with respect to frequency hopping systems is the hopping rate r_h that determines the bandwidth covered. There are two basic frequency hopping systems including slow and fast frequency hoppers. Slow frequency hoppers are systems where the symbol rate r_s is a multiple of r_h, or in other words several symbols are transmitted in a hop. On the other hand, in fast frequency hoppers, r_h is a multiple of r_s, that is one symbol is transmitted over multiple hops. As in the context of spread spectrum, a chip in the context of frequency hopping is the shortest uninterrupted waveform in the system. Therefore, for fast hoppers a chip is the waveform transmitted during the hop, while for slow hoppers a chip is a data symbol. The processing gain for frequency hopping systems is defined as follows:

$$G = \frac{W}{r_b} \tag{2.16}$$

where W represents the frequency bandwidth and r_b is the bit rate. The Bluetooth [1] technology and one of the PHY layers for the IEEE 802.11 [4] specifications use (slow) frequency hopping.

2.2   Media access control layer

The media access control (MAC) protocol consists of a set of rules that regulate the sharing and access of a communication channel as shown in Figure 2.4. In wireless communications, radio waves represent the communication channel that must be shared among multiple users. Like all multiple access environments, only a single device can successfully transmit a message at any given time. Simultaneous transmissions result in interference and the collision of the messages at the receiver. In this section, we discuss some of the best known channel partitioning mechanisms and access control protocols.

2.2.1   Channel partitioning

Channel sharing techniques are broken up into three major categories depending on how they partition and make use of the channel: (1) time division multiple access (TDMA), (2) frequency division multiple access (FDMA), (3) code division multiple access (CDMA).

Time division multiple access

In TDMA, the channel is partitioned into time intervals, and each user is allowed to use the full transmission capacity of the channel for some amount of time, as illustrated in Figure 2.5. Only a single user is allowed to transmit at a time, therefore there are no collisions of packets at the receiver. Time is divided into slots and each user is given a certain number of slots every so often. This is better known as scheduling and will be discussed in more detail in the scheduling section of this chapter. TDMA is often used when traffic conditions remain constant and delay variation requirements need to be kept low. Although slot allocation can be adaptive,

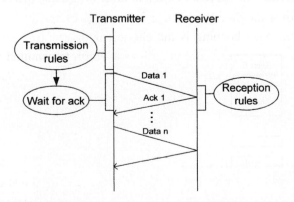

Fig. 2.4. MAC transmission and reception rules.

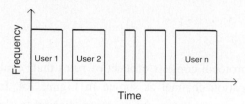

Fig. 2.5. Time division multiple access.

TDMA can result in the underutilization of the channel in cases when users do not have enough data to send. A hybrid TDMA/FDMA system was used for the global system for mobile communications (GSM) where a 25 MHz band is divided first into 200 kHz channels. Each station can use one or more channels in its cell. A channel is further divided into 120 ms multiframes that are further divided into twenty-six frames that contain eight slots each.

Frequency division multiple access

In FDMA, users are separated in the frequency domain. The channel is divided into a number of frequency slots, each of which accommodates the signal of an individual user, as shown in Figure 2.6. Each frequency slot is assigned to a single user that uses a different modulation to place a signal in the appropriate slot. The combined signal is transmitted and the receiver recovers the signal destined for it by use of appropriate demodulation. Since in FDMA a transmitter uses a fixed portion of the frequency band all the time, this method is most suitable for constant bit rate and voice applications. This results in efficiencies when traffic is bursty. FDMA was used in the first generation advanced mobile phone system (AMPS) where a 50 MHz channel was broken up into 30 kHz transmission channels for carrying voice traffic resulting in 832 two-way channels.

Orthogonal frequency division multiplexing (OFDM) is another form of frequency division multiplexing that divides the spectrum into a number of equally spaced tones

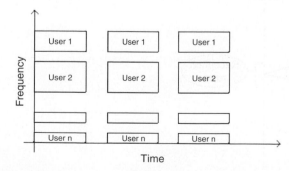

Fig. 2.6. Frequency division multiple access.

and carries a portion of a user's information on each tone. In OFDM each tone is orthogonal with every other tone. While FDM typically requires there to be frequency guard bands between the frequencies so that they do not interfere with each other, OFDM allows the spectrum of each tone to overlap. Because the tones are orthogonal, they do not interfere with each other and the overall amount of spectrum required is reduced. The benefits of OFDM are high spectral efficiency, resiliency to RF interference, and lower multipath distortion. OFDM is used in high speed WLAN [8] and WMAN [14] systems.

Code division multiple access

CDMA is another type of channelization. In TDMA or FDMA, different user transmissions are clearly separated in either time or frequency, respectively. In CDMA the transmissions from different users occupy the entire frequency band at the same time. The transmissions are separated by the fact that different codes are used to produce different signals from different users. The receiver uses these codes to recover the transmitted signal destined for it, as shown in Figure 2.7. Each user's code consists of a unique binary pseudo-random sequence of chip values (represented by "+1" and "−1") and is produced by a special code that appears to be random and repeats after a long period of time. The user information bits are multiplied by the user's code and then digitally modulated and transmitted over the medium. Since codes assigned to different users are orthogonal to one another, the interference between two users sharing the same bandwidth simultaneously is small. Still, there is a limitation on the maximum number of users possible before exceeding an acceptable interference threshold. Although the spectrum efficiency factor is 1, as the number of users increases, the interference and the resulting bit error rate at the receiver gradually increases as well. Additional techniques to change the code length and transmitted power dynamically exploit traffic characteristics and user activity for added efficiency.

Due to cost and complexity associated with its implementation, including real-time pseudo-random code assignment and GPS synchronization, CDMA is mostly

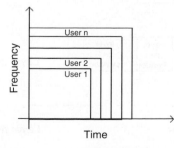

Fig. 2.7. Code division multiple access.

applied to wide area networks and licensed band operation. CDMA is used in the third generation of GSM, also known as the universal mobile telecommunications services (UMTS).

2.2.2   Access control protocol

Access control protocols are used in conjunction with channel partitioning techniques in order to allow for the efficient sharing of the medium. The basic operation of an access control protocol is as follows. A packet arriving at the MAC layer from the application layer first undergoes segmentation. It is then packaged into MAC layer frames by adding specific header and trailer information and filling the MAC payload frame. Transmission rules are then applied in order to send the packet to its destination. Since in wireless systems collisions at the receiver cannot be detected by the sender, the status of the reception has to be communicated back to the sender in the form of an acknowledgment message. In case of an error in the reception, an error correction scheme may be applied in order to recover the packet. Otherwise, the sender is notified to perform a retransmission. At that point, the sender enters a collision resolution phase in order to ensure a higher probability of success when retransmission is attempted. Bandwidth allocation schemes and scheduling schemes are used to provide fairness either among different users or different traffic types per user.

Based on their functionality, several key components of access control protocols are identified and discussed in more detail below.

Packetization

The segmentation of payload data units (PDU) coming from the higher layer is often required since each MAC layer protocol has its own frame size and structure, as illustrated in Figure 2.8. In the IEEE 802 protocol stack, the MAC layer protocol

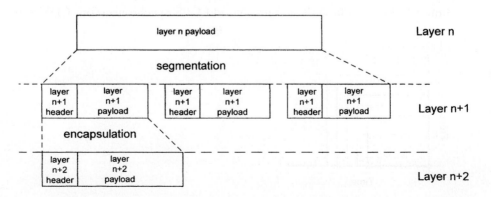

Fig. 2.8. Packet segmentation and encapsulation.

receives packets from the logical link control (LLC) layer that is directly above. However, not all technologies use this layering; for example, Bluetooth and Zigbee do not share this layering structure. Each technology also specifies one or several MAC frame structures depending on the packet type, whether it is a control or data packet, and the type of traffic it is intended to carry. In addition, different packet sizes can be chosen in order to control burstiness and channel errors, and to comply with specific allocation rules.

Scheduling

The main objectives in scheduling algorithms is to guarantee the fair servicing of users and the distribution of shared resources. In the case of radio communication, scheduling is directly linked to transmission rules and it can be generally sub-divided into two broad categories, namely reservation-based, and "free-for-all" or contention-based access.

In reservation-based systems, such as that illustrated in Figure 2.9, transmission rights are assigned to each user in turn so as to avoid simultaneous transmissions and packet collisions. Reservation systems are typically associated with TDMA channel partitioning, and the bandwidth is either pre-allocated or dynamically allocated. In the case of pre-allocation, a scheduling algorithm determines how many slots each user gets and how often according to its traffic requirements. In the dynamic allocation, users may request bandwidth on demand through a request–grant message exchange or a periodic poll mechanism. It is observed that reservation systems are associated with one-to-many communications where one node, typically a base station, is responsible for dividing up the bandwidth. Like TDMA channel partitioning, reservation-based systems may suffer from underutilization of the channel in cases when too many users are idle.

In contention-based systems, such as that shown in Figure 2.10, these inefficiencies are overcome by giving the rights to transmit to all users who wish to transmit and hoping that they will not transmit at the same time. Contention stems from the fact that two or more users may decide to use the channel at the same time.

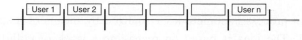

Fig. 2.9. Reservation-based access.

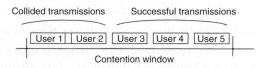

Fig. 2.10. Contention-based access.

Both mechanisms can be combined in order to increase further the throughput of the channel, where short packets are used in contention to reserve long non-contending slots for sending data. Thus, only short slots are wasted by idles or collisions, leading to a better overall channel efficiency.

Collision resolution

In the case where a form of contention is used, a mechanism is needed to resolve collisions and control the retransmission of packets. Many strategies have been developed to solve the generic problem of having two or more transmitters sending packets simultaneously. They can be divided into two basic paradigms. One is based on random retransmissions, like the one used for the Aloha network, which was the first radio packet network developed to link the various campuses at the University of Hawaii in the early 1970s. Nodes in the Aloha network attempt to retransmit collided messages hoping for no interference from other nodes [64]. Several schemes have been developed since the original Aloha scheme in various attempts to improve its performance [74].

One variation, known as p-persistence, associates with each slot a transmission probability, p, where collided packets are retransmitted with a probability p. This probability can be controlled by a central node that knows more about the state of the network. It can also be based on a backoff period computed locally in order to ensure that not all collided nodes will retransmit their packets at the same time. The so-called binary exponential backoff mechanism doubles the backoff window every time a collision occurs. A station randomly chooses a time within this backoff window when scheduling a retransmission. This process is repeated until a request is successfully received at the destination.

The other paradigm consists of splitting collided nodes into a tree structure. In this tree-based mechanism, all nodes involved in a collision split into a number of subsets. The first subset transmits first, followed by the second subset, then the remaining subsets. The chances of future collisions are reduced by forcing users that collided in the same slot (assuming a slotted frame structure) to retransmit their packets in different slots in the future, thus distributing the number of contending stations over several slots.

The collision resolution process is sometimes visualized by means of a stack, as illustrated in Figure 2.11. The idea is based on managing a stack of collided and newcomer users, where each stack level corresponds to a different group of users ranked from top to bottom. Note that the stack in this case is only a visualization and need not be implemented. Although the stack is a good visual aid for understanding the dynamics of splitting algorithms and is usually associated with tree-based contention resolution mechanisms [30], the idea can be applied to p-persistence schemes as well.

For n-tree algorithms, colliding transmissions are split into n groups and each group is retransmitted in a different slot. The process repeats until all collisions are

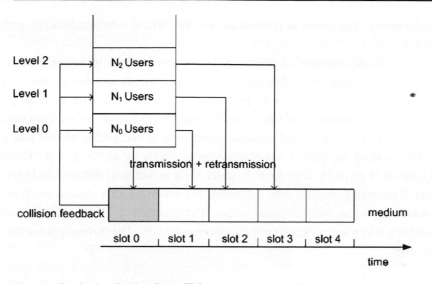

Fig. 2.11. Stack visualization for collision resolution algorithms.

resolved. Assuming that the next group of users that are allowed to transmit occupies the bottom level, collided users entering the stack may either go to the top or the bottom of the stack, thus achieving a FIFO or a LIFO ordering, respectively. Applying the stack management visualization to p-persistence schemes leads to a two-level stack. Based on a collision feedback, users at levels 0 and 1 remain at the same level with a probability p and $1 - p$, respectively. On the other hand, users at levels 0 and 1 change levels with probability $1 - p$ and p, respectively. After a successful feedback, stations at level 1 either go to level 0 with probability p or remain at level 1 with probability $1 - p$.

It is evident from the stack visualization that as the network load increases, the delay variation for the p-persistent collision resolution algorithm is often much higher than for the tree-based algorithms. This result is expected considering the random nature of the p-persistence.

These two basic schemes have evolved to adapt to various network environments and constraints. In radio environments, the ratio of propagation time to transmission delay is relatively small ($\ll 1$), and therefore stations have the ability to monitor the transmission channel before sending a message. If the medium is sensed as busy, the packet transmission is postponed until a later time, when the medium becomes idle, thus avoiding a collision and the need for an almost certain retransmission. This is the so-called carrier sense multiple access scheme with collision avoidance (CSMA/CA) widely used in wireless networks.

Obviously slight variations in the collision resolution scheme and the parameters that control the access to the medium lead to significant differences in the performance obtained. In order to illustrate this concept, let us consider the performance of the Aloha scheme since it is the simplest to analyse. Others have been also analysed in the literature. The reader is referred to refs [69,74] for a comprehensive analysis on the topic.

Let N be the number of nodes in the system, and let λ be the packet arrival distribution at each node. The aggregate packet arrival rate for all the contending nodes in the system is given by $\rho = N\lambda m$, where m is the size of each packet in units of time. However, because simultaneous packet transmissions from contending nodes lead to collisions and retransmissions, the actual packet arrival rate on each node including the packet retransmissions is given by κ, for $\kappa > \lambda$. The channel utilization is given by $G = N\kappa m$. In order for a message of duration m to be sucessfully transmitted without any collisions, no other packets should be sent on the channel during an interval equal to $2m$. The probability of no collision during $2m$ assuming a Poisson packet arrival distribution is e^{-2G}. The throughput for the system becomes

$$S = Ge^{-2G} \tag{2.17}$$

Observe that the throughput for the Aloha system is simply doubled when slot boundaries are defined and each node has to wait until the beginning of each slot in order to transmit its packet. This resulting system is known as slotted Aloha and its throughput is equal to Ge^{-G}. It easy to observe that both of these systems reach their peak when $G = 1$. Thus $S = 1/2e$ and $S = 1/e$ for Aloha and slotted Aloha, respectively.

Error detection and correction

Whether it is due to packet collisions or noise, wireless transmission systems can experience error rates as high as 10^{-3} or worse. Therefore, error detection and recovery become necessary.

The basic idea in error detection is as follows. The transmitter uses an encoder to insert additional redundancy bits to the information bits to be transmitted. These redundant bits together with the information bits form a so-called code. Then the receiver uses a decoder in order to check the code found in the bits received against a pre-determined code. If they match, the information is error-free. Otherwise there may be errors in the information received.

The simplest form of error detection is the parity check code that takes k information bits and appends a single check bit to form a codeword. The parity ensures that the total number of ones in the codeword is even. Other schemes include checksums and polynomial codes (used in cyclic redundancy check) that involve manipulating the information to be sent in order to compute a code. Linear block codes including Hamming codes, BCH, and Read–Solomon are extremely popular. Convolutional codes are another important class of error detection and correction codes. The main difference between linear codes and convolutional codes is that convolutional codes operate on overlapping blocks of information whereas linear codes use non-overlapping blocks of information.

There are two basic approches to error recovery. The first involves forward error correction (FEC), which consists of an attempt to correct errors after they are detected. FEC makes use of additional redundancy bits in the information sent in order to detect and recover the original information sent. The success of FEC depends on the error characteristics and location in the message. FEC is generally used in environments where a return channel is not always available or the feedback delay is long, for example in satellite or deep space communications.

The second approach involves the use of an automatic repeat request (ARQ) after errors are detected. ARQ can be in the form of an acknowledgment message sent back to the transmitter about the status of the packet reception at the destination. It can also be implied or "piggybacked" in a return packet from the destination to the source. While FEC uses extra bits for redundancy, ARQ wastes additional bandwidth and introduces extra feedback delay. Thus the suitability of each technique depends on the application and the environment they are used in. FEC is more common in applications with stringent delay requirements such as voice and video where retransmission delays are not tolerable. On the other hand, ARQ is more common for data applications.

Whether ARQ is used or not, error detection and correction are typically performed at the same time. The choice of code determines whether correction is carried out in addition to detection. So, what are useful performance metrics to assist in evaluating the effectiveness of error detection and correction codes?

Redundancy is probably the first metric to consider. It consists of the amount of bits that are transmitted in addition to the information bits. Therefore, if k bits of information are encoded with a total of n bits, the code (n, k) has a redundancy ratio of $(n - k)/k$ and a code rate of k/n. Another common metric to consider includes the probability of detection failure, also known as the probability of undetected errors. Basically, given an error probability p in the bit stream to be decoded, this is the probability that the error detection scheme fails to detect the errors present in the bit stream. For example, the single parity check code probability of detection failure is equal to the probablity of error patterns with an even number of ones, since these errors are undetectable. When calculating the probability of undetected errors, one has to make assumptions about the error distribution, or, in other words, one must consider how many errors are present and where they are located in the bit stream. It is observed that the probability of successfully detecting errors generally depends on the error characteristics, including randomness and burstiness. However, most estimations of the probability of error detection failure assume that errors are uncorrelated, which is not realistic. This is especially true in the case of interference, since errors occur in a bit stream due to the collision of two packets in time. Therefore the probability of having two consecutive errors is much higher than in the case of noise.

Other performance metrics to consider are related to a code structure, since it is the structure that determines the code's effectiveness. Thus, the Hamming weight, $w(c)$, of a code c is defined as the number of non-zero elements it contains. Also, the Hamming distance, d, between two codes $c1$ and $c2$ is defined as the number of elements in which they differ. Observe that $d(c1, d2) = w(c1 + c2)$; d_{min} is the minimum distance between all pairs of codes defined in a space S. The error detecting and correcting capabilities are often related to d_{min}. The maximum number of correctable errors, t, is found to be equal to [75]

$$t = \lfloor \frac{d_{min} - 1}{2} \rfloor \tag{2.18}$$

In addition, the number of errors detected, e, is given by

$$e = d_{min} - 1 \tag{2.19}$$

2.2.3   Key design factors

There are many challenges specific to wireless environments that come into play when designing a MAC protocol. Key factors include network topology, power, duplexity, coverage area, and sensing limitations.

(i) Network topology – there are in wireless communications a number of network topologies envisaged ranging in capabilities and resources. Nodes may be moving, so the topology may be changing as well. Network topology is generally described in terms of a hierarchy. Centralized topologies are those that include a node that plays the role of a base station and controls the use of the network. On the other hand, ad hoc networks are those where nodes communicate peer-to-peer without the need of a central node.

(ii) Power – reducing power consumption is extremly desirable in wireless communications due to the need to save battery life. In addition, limiting the transmit power has the added benefit of reducing the level of signal interference with neighboring devices. This becomes a significant factor in the design of a MAC protocol having the objective of minimizing packet collisions and retransmissions.

(iii) Coverage area – signal strength decays according to a power law with distance due to channel propgation properties. Therefore, radio transmission covers only a certain radius for a given transmission power, channel, and signal type.

(iv) Duplexity – in radio systems it is not always possible to send and receive simultaneously unless multiple transmitters and receivers are available within the same system. This is due to the fact that the signal energy for the transmission leaks in the receive path. This self-interference signal is generally much

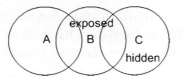

Fig. 2.12. Exposed and hidden nodes.

higher than the received signal, which makes it impossible to receive and detect collisions while transmitting. Therefore collision detection schemes cannot be used during transmission (as in Ethernet), and collision avoidance schemes have to be used instead. Collision avoidance schemes include medium sensing and delayed access or backoff.

(v) Channel errors – errors in the wireless channels are mainly due to time-varying radio propagation properties and low signal strength. Signal propagation properties, including reflection, diffraction, and scattering, break the transmitted signal in a number of superimposed signals at the receiver that are time-shifted and attenuated. Coherence time determines the rate of channel variations when the signal strength drops below 3 dB. Fading occurs when the received signal strength drops below a certain threshold.

(vi) Sensing limitations – the ability of the sender to sense the channel and detect transmissions is generally limited by the proximity of the sender and its location with respect to the receiver. Only nodes within a specific radius of the transmitter can detect activity in the channel. This gives rise to the hidden node and exposed node problems. A hidden node is one that is within the range of the intended destination but out of range of the sender. An exposed node is one that is within the range of the sender but out of range of the destination (see Figure 2.12).

2.3   Examples of wireless protocols

In this section we briefly overview two widely available wireless protocols, namely the Bluetooth specifications [1] and IEEE 802.11b [4]. The discussion is largely based on the terms and concepts introduced earlier in this chapter.

2.3.1   Bluetooth

The Bluetooth technology [1] is a short range (0–10 m) wireless link technology aimed at replacing non-interoperable proprietary cables that connect phones, laptops, PDAs and other portable devices. Bluetooth operates in the ISM frequency band starting at 2.402 GHz and ending at 2.480 GHz in the USA and Europe; 79 RF channels of 1 MHz width are defined. The air interface is based on an antenna power of 1 mW. The signal is modulated using binary Gaussian frequency shift keying (GFSK).

The raw data rate is defined at 1 Mbit/s. A time division multiplexing (TDM) technique divides the channel into 625 μs slots resulting in 1600 slots/s. Transmission occurs in packets that occupy an odd number of slots (one, three, or five). Each packet is transmitted on a different hop frequency with a maximum frequency hopping rate of 1600 hops/s for packets occupying a single slot, and a minimum hopping rate of 320 hops/s for packets occupying five slots. Note that every slot has a frequency associated with it; however, transmission of a packet occupying multiple slots always uses the frequency associated with the first slot.

Two or more units communicating on the same channel form a piconet, where one unit operates as a master and the others (a maximum of seven active at the same time) act as slaves. A channel is defined as a unique pseudo-random frequency hopping sequence derived from the master device's 48-bit address and its Bluetooth clock value. The algorithm for generating the frequency hopping pattern works as follows. Frequencies are sorted into a list of even and odd frequencies in the 2.402–2.480 GHz range. A window consisting of the first thirty-two frequencies in the sorted list is chosen. After all thirty-two frequencies in that window are visited once in a random order, a new window is set including sixteen frequencies of the previous window and sixteen new frequencies in the sorted list. Slaves in the piconet synchronize their timing and frequency hopping to the master upon connection establishment. In the connection mode, the master controls the access to the channel using a polling scheme where master and slave transmissions alternate. A slave packet always follows a

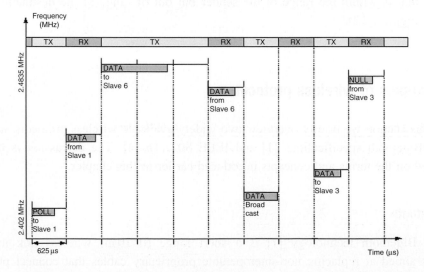

Fig. 2.13. Master TX/RX hopping sequence (Figure appeared in ref. [44]).

master packet transmission, as illustrated in Figure 2.13, which depicts the master's view of the slotted TX/RX channel. A slave needs to respond to a master's packet that is specifically addressed to it. If it does not have any data to send, it sends a NULL packet.

Depending on whether a connection carries data or voice traffic, two types of link connections between the master and the slave are envisaged.

The asynchronous connection link (ACL) is an asymmetric point-to-point connection for transporting data packets between the master and the slave. An automatic repeat request (ARQ) procedure is applied to ACL packets where packets are retransmitted in case of loss until a positive acknowledgment (ACK) is received at the source. Several packet formats are defined for the ACL link, namely, DMn and DHn packets that occupy an odd number of slots ($n = 1, 3, 5$ slots). The DM format includes FEC, while the DH format does not.

The synchronous connection-oriented (SCO) link is a symmetric point-to-point voice connection between a master and a slave where packets are sent at regular time intervals, denoted by T_{SCO} time slots. The slave responds to the SCO packets sent by the master in the slot following the master's transmission. The rate for the bit stream in each direction is set to 64 kbit/s and determines the value of T_{SCO} for different packet formats. It is set to either two, four or six time slots for HV1, HV2, or HV3 packet formats, respectively. HV1 contains 80 bits of information transmitted every two slots or 1250 μs. Similarly, HV2 and HV3 contain 160 and 240 bits of information and are transmitted every 2500 μs and 3750 μs, respectively. All three formats of SCO packets are never retransmitted in the case of packet loss or error.

Both ACL and SCO packets have the same packet format, as shown in Figure 2.14. It consists of a 72-bit access code used for message identification and synchronization, a 54-bit header and a variable length payload. The header includes the active slave address, the packet type, and a header error check (HEC). There is also the ARQN bit that is set to either 1 or 0 depending on whether the previous packet was successfully

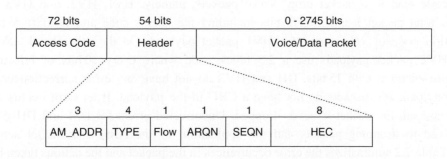

Fig. 2.14. Bluetooth packet format.

Table 2.1. *Bluetooth packet types*

Link	Packet	Payload (bytes)	Payload FEC code rate	Burst length (μs)	Occupied slots
Control	NULL	0		126	1
	POLL	0		126	1
ACL	DM1	0–17	2/3	171–366	1
	DM3	0–121	2/3	186–1626	3
	DM5	0–224	2/3	186–2871	5
	DH1	0–27	none	150–366	1
	DH3	0–183	none	158–1622	3
	DH5	0–339	none	158–2870	5
SCO	HV1	10	1/3	366	1
	HV2	20	2/3	366	1
	HV3	30	none	366	1
	DV	10 for voice + (0–9) for data	2/3 for data	238–356	1

received or not, and the sequence number (SEQN) bit used to provide a sequential ordering of data packets and to filter out retransmissions at the destination. The payload contains either a voice or a data packet depending on the type of link connections that are established between a master and a slave.

Table 2.1 summarizes the various packet types available for the SCO, ACL and link control packets such as POLL (used by the master) and NULL (used by either the master or slave device).

A repetition code of 1/3 (i.e. every bit of information is repeated three times) is applied to the header, and a block code with minimum distance $d_{min} = 14$ is applied to the access code so that up to thirteen errors are detected and $\lfloor (d_{min} - 1)/2 \rfloor = 6$ errors can be corrected. Note that uncorrected errors in the header and the access code lead to a packet drop. Voice packets, namely, HV1, HV2, and HV3, have a total packet length of 366 bits including the access code and header. A repetition code of 1/3 is used for HV1 packet payload. On the other hand, DM and HV2 packet payloads use a 2/3 block code where every 10 bits of information are encoded with 15 bits. DH and HV3 do not have any error correction on their payload. HV packets do not have a CRC in the payload. If an error occurs in the payload, the packet is never dropped. Uncorrected errors for DM and DH packets lead to dropping packets and the application of the ARQ and SEQN schemes. Table 2.2 summarizes the error occurrences in the packet and the actions taken by the protocol.

Table 2.2. *Summary of error occurrences in the packet and actions taken when errors are not corrected*

Error location	Error correction	Action taken
Access code	$d_{min} = 14$	packet is dropped
Packet header	1/3 repetition	packet is dropped
HV1 payload	1/3 repetition	packet is accepted
HV2 payload	2/3 block code	packet is accepted
HV3 payload	no FEC	packet is accepted
DM1, DM3, DM5 payload	2/3 block code	packet is dropped
DH1, DH3, DH5 payload	no FEC	packet is accepted

2.3.2   IEEE 802.11b

The IEEE 802.11 standard [4] defines both the physical (PHY) and medium access control (MAC) layer protocols for WLANs. From now on, we will be using WLAN and 802.11 interchangeably.

The IEEE 802.11 standard calls for three different PHY specifications: frequency hopping (FH) spread spectrum, direct sequence spread spectrum (DS), and infrared (IR). The transmit power for DS and FH devices is defined at a maximum of 1 W.

Under FH, each station's signal hops from one modulating frequency to another in a pre-determined pseudo-random sequence. Both transmitting and receiving stations are synchronized and follow the same frequency sequence. There are 79 channels defined in the (2.4000–2.4835) GHz region spaced 1 MHz apart. The minimum hop rate is 2.5 hops/s and the maximum is unspecified. The basic access rates of 1 and 2 Mbit/s use multilevel Gaussian frequency shift keying (GFSK).

A DS transmitter converts the data stream into a symbol stream, where each symbol represents a group of multiple bits to spread over a relatively wideband channel of 22 MHz. The basic data rate is 1 Mbit/s encoded with differential binary phase shift keying (DBPSK) or 2 Mbit/s using differential quadrature phase shift keying (DQPSK). Higher rates of 5.5 and 11 Mbit/s are also available with techniques combining pulse-position-modulation (PPM) and quadrature amplitude modulation (QAM).

The IEEE 802.11 MAC layer specifications are common to all PHYs and data rates, and coordinate the communication between stations and control the behavior of users who want to access the network. The distributed coordination function (DCF), which describes the default MAC protocol operation, is based on a scheme known as carrier sense, multiple access, collision avoidance (CSMA/CA). Both the MAC and PHY layers cooperate in order to implement collision avoidance procedures. The PHY layer samples the received energy over the medium transmitting data and uses a clear channel assessment (CCA) algorithm to determine if the channel is clear.

This is accomplished by measuring the RF energy at the antenna and determining the strength of the received signal commonly known as RSSI, or received signal strength indicator. In addition, carrier sense can be used to determine if the channel is available. This technique is more selective since it verifies that the signal is the same carrier type as 802.11 transmitters.

A virtual carrier sense mechanism is also provided at the MAC layer. It uses the request-to-send (RTS) and clear-to-send (CTS) message exchange to make predictions of future traffic on the medium and updates the network allocation vector (NAV) available in stations. Communication is established when one of the wireless nodes sends a short RTS frame. The receiving station issues a CTS frame that echoes the sender's address. If the CTS frame is not received, it is assumed that a collision occurred, and the RTS process starts over.

Regardless of whether the virtual carrier sense routine is used or not, the MAC is required to implement a basic access procedure (depicted in Figure 2.15) as follows. If a station has data to send, it waits for the channel to be idle through the use of the CSMA/CA algorithm. If the medium is sensed idle for a period greater than a DCF interframe space (DIFS), the station goes into a backoff procedure before it sends its frame. Upon the successful reception of a frame, the destination station returns an ACK frame after a short interframe space (SIFS). The backoff window is based on a random value uniformly distributed in the interval $[0, CW]$, where CW represents the contention window parameter and is varied between CW_{min} and CW_{max}. If the medium is determined busy at any time during the backoff slot, the backoff procedure is suspended. It is resumed after the medium has been idle for the duration of the DIFS period. If an ACK is not received within an ACK timeout interval, the station assumes that either the data frame or the ACK was lost and needs to retransmit its data frame by repeating the basic access procedure.

The standard specifies an overall MAC frame format as depicted in Figure 2.16. Regardless of the PHY layer specifications used and the bit rates specified, the MAC header is always sent at 1 Mbit/s. The frame control field carries the type of frame

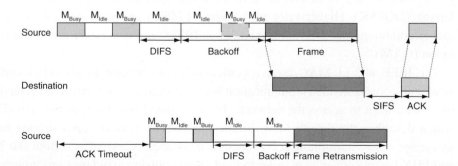

Fig. 2.15. WLAN frame transmission scheme (taken from ref. [44]). (a) Successful frame transmission. (b) Frane retransmission. M_{Busy} = medium is busy; M_{Idle} = medium is idle.

2 bytes	2 bytes	6 bytes	6 bytes	6 bytes	2 bytes	6 bytes	0-2312 bytes	4 bytes
Frame Control	Duration /ID	Address 1	Address 2	Address 3	Seq. Control	Address 4	Frame Body	FCS

Fig. 2.16. IEEE 802.11 frame format.

being sent. There are three frame types defined, namely the management, control, and data frames. Management frames include beacons, probe request and probe response frames. Control frames include power-save, request-to-send (RTS), clear-to-Send (CTS), and acknowledgment (ACK) frames. The duration field contains information on the duration of the next frame transmission. The address fields contain the source, destination, transmitting and receiving stations IEEE MAC 48-bit addresses. The transmitting and receiving addresses identify the intermediate stations involved in the transmission and reception of the frame. The sequence control field indicates the sequence number of the frame being transmitted and the frame check sequence (FCS) field contains the result of a cyclic redundancy check (CRC) algorithm applied on the MAC header and frame body. At least one bit error causes the packet to be dropped and retransmitted.

3 Interference performance evaluation

Our objectives in this chapter are to describe the basic building blocks in performance evaluation as we focus on identifying and understanding the effects of interference in wireless communications and its impact on system performance.

Since we set out to evaluate the effects of interference on performance, the first question we ask is what is interference? The term "interference" has been extensively used in the context of communication, in both wired and wireless systems. While an accurate definition may be dependent on the specifics of the context considered, the term generally refers to signal impairments due to factors in the environment such as channel propagation properties, other radiated power, and noise.

The second question is concerned with the performance evaluation of interference, namely, what are the quantitative measures that characterize interference, and consequently how should the resulting level of performance be quantified? One interference metric that has been used extensively includes the so-called signal to interference ratio. However, this measure does not characterize completely the resulting performance since performance is often tied to the quality of service requirements, which vary depending on the application considered. Our objective is to provide a list of performance metrics that can accurately quantify the network performance from an application perspective.

Since not all systems behave in the same way given the same level of interference, an important aspect of performance evaluation is to identify parameters that impact performance. This is crucial for understanding the behavior of systems and even more important for designing better systems and solutions to mitigate interference. As we shall see, most wireless protocol components, for example signal modulation, receiver detection, contention resolution, as discussed in Chapter 2, have a direct effect on interference and the resulting performance. In this chapter, we list several parameters and present examples in order to highlight the effects of the parameters selected.

The remainder of this chapter is organized as follows. First, we present a definition for interference in the context of wireless systems. Secondly, we discuss several metrics that can be used to quantify the effects of interference on performance.

Thirdly, we identify protocol parameters that impact interference and the resulting system performance.

3.1   Interference model

Interference in the context of wireless systems usually refers to either one of the following two definitions: (1) multiple (more than two) simultaneous packet transmissions causing packets to collide at the receiver, (2) physical factors in the radio propagation channel.

There are many effective techniques for dealing with simultaneous transmissions, such as carrier sense multiple access mechanisms devised to limit the effects of collisions resulting from simultaneous transmissions. While some of these techniques may be effective in reducing the total number of collisions incurred, there are some collisions that are unavoidable, due to what is known as the "hidden" node problem [64] where station A, believing the medium is idle because it cannot hear station B, begins transmission, while station B is in the middle of a transmission.

Interference caused by the radio wave propagation properties in the air has been extensively studied in the classical communication literature [68,73,77]. Various physical impairments, such as multipath fading, contribute to errors in the wireless channel. Multipath fading can be characterized by modeling the received signal as a sum of copies of the transmitted waveform, each with a random amplitude and delay (phase) [50].

In the context of this chapter, and of this book in general, we focus exclusively on the form of interference caused by multiple simultaneous packet transmissions. Furthermore, we define an interference model to consist of at least two heterogeneous communication systems as illustrated in Figure 3.1. A communication system consists

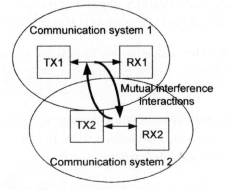

Fig. 3.1. Mutual interference between communication systems.

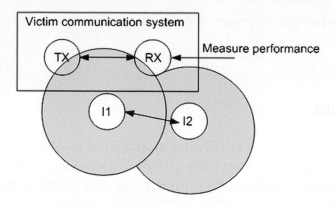

Fig. 3.2. Measuring interference.

of a transmitter and a receiver. There are two interfering devices also shown in the figure. The assumption here is that the communication systems do not communicate directly since they may be implementing different protocols. Due to closed-loop interactions between the interfering and the victim systems, each system is both a victim and an interfering system. In the case where the system under study is subject to interference, this system represents the victim. If, on the other hand, it is considered to cause interference, it represents the interferer system. We usually consider the effects of interference on the victim system performance. The victim system performance is quantified at the receiver node, as shown in Figure 3.2.

3.2 Performance metrics

In this section we overview some of the performance metrics used in the evaluation of a communication system and more specifically in the evaluation of interference. These measurements are broken up into two categories depending on whether they characterize the performance of the system at the PHY layer or at higher layers (MAC and above).

3.2.1 PHY layer performance measures

The basic figure of merit for communication systems is the average signal power to average noise power ratio, S/N or SNR, measured in decibels (dB). In fact, this ratio of two powers constitutes the definition of a decibel, where

$$1\,\mathrm{dB} = 10\log_{10}\frac{P_1}{P_2} \qquad (3.1)$$

Note that 1 dBm or dBmW represents a measured power level in decibels relative to 1 mW; i.e.

$$1\,\text{dBm} = 10\log_{10}\frac{1.25983\,\text{mW}}{1\,\text{mW}} \tag{3.2}$$

or equivalently

$$1\,\text{dBm} = \frac{1\,\text{dB}}{1\,\text{mW}} \tag{3.3}$$

Note that this a theoretical measure that is computed based on estimates for the signal power level and the noise level. In digital communications, SNR is often replaced by E_b/N_0, a normalized version of SNR, where E_b is the bit energy and is equal to the signal power S times the time to transmit one bit T_b. In addition, N_0 is the noise power spectral density and can be expressed in terms of the noise power, N, divided by the bandwidth, W. If R_b is the bit rate, T_b is equal to $1/R_b$. Thus,

$$\frac{E_b}{N_0} = \frac{ST_b}{N/W} = \frac{S/R_b}{N/W} \tag{3.4}$$

Also, by rearranging,

$$\frac{E_b}{N_0} = \frac{S}{N}\left(\frac{W}{R}\right) \tag{3.5}$$

One of the most important metrics of performance in digital communications is a plot of the probability of bit error P_b versus E_b/N_0, as shown in Figure 3.3, which represents the general waterfall-like shape of most such curves that characterize communication systems.

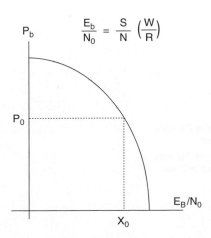

Fig. 3.3. General shape of P_b versus E_b/N_0.

For $E_b/N_0 \geq x_0$, $P_b \leq P_0$; i.e. the smaller the ratio E_b/N_0 the higher the probability of error. This curve depends on the modulation chosen and the receiver detection design. Thus, E_b/N_0 can be seen as a required ratio in order to obtain a specific bit error performance.

It is important to observe here that the probability of bit error is a theoretical measure of performance and that in real systems it is impossible to measure since the receiver can only estimate the signal transmitted and does not have a copy of the transmitted signal. In simulation modeling, a copy of the transmitted signal is compared to the received signal and a bit error rate can be computed.

3.2.2   Higher layer performance measures

The measures defined in this section can apply to almost all layers above the PHY layer. In defining these metrics, it is important to specify what needs to be transmitted including packet format and type and at what layer the measurements will be conducted in order to define completely the communication protocols involved in the transmission and reception procedures.

Most importantly, performance evaluation is concerned with quantifying, i.e. with how many packets make it successfully to the other end and how long does it take then to reach the other end. Essentially, the characterization of a system is based on detailed accounting to track how many packets are transmitted, received, and lost, as shown in Figure 3.4, in addition to timing information. Typically, most implementations include the following metrics.

(i) Packet loss – a measure of how many packets are correctly received at the other end. It is calculated as the ratio of packets dropped at the receiver due to reception errors over the total number of packets transmitted.

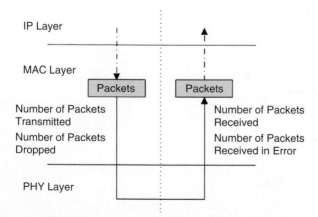

Fig. 3.4. Higher layer performance measures.

(ii) Delay in seconds – a measure of the time it takes a packet to be transmitted and received at the other end. The delay measure can be further broken up into several components, including queueing delay or the time the packet spent in the transmitter's queue before being transmitted; transmission delay, which depends on the transmission's bit rate and the length of the packet; access delay, which includes the backoff and scheduling time before a packet is transmitted on the medium. The retransmission delay accounts for the time to retransmit the packet if it is lost.

(iii) Throughput in bit/s – measures the number of bits (or packets) received divided by the time taken to complete their transmission. Goodput is the measure of bits received excluding any overhead bits in terms of retransmissions and/or packet headers. Efficiency is another measure of how many useful packets are received and forwarded to the receiver's higher layer divided by the total number of packets transmitted. This measure accounts for retransmissions and errors.

It is important to note that all of the metrics listed can be defined at any layer above the PHY layer. The packet loss, for example, can be implemented at any layer above the PHY layer, although it is mostly applicable to the MAC layer. Here the probability of PHY layer bit errors is applied to a packet being received and translates into a packet error rate, or the probability of finding an error in the packet. Most packets include a parity check sum that enables the detection and sometimes correction of errors found in packets. Depending on the location and number of errors and whether an error correction code is used, errors found may be corrected before a decision is made to accept or reject the packet. If the check sum computed is different from the one found in the packet, the packet is discarded. This constitutes the packet loss measure, which represents the main measure used in the system level performance in order to quantify interference. If no error correction code is applied on the packet, a single error found in the packet leads to the packet being dropped or lost at the receiver. While the packet error rate represents errors in packets before error correction codes are applied, the packet loss is measured after performing error correction. Thus, this measure depends not only on the SNR, the error correction efficiency, but also on the time, frequency overlap, and the traffic distribution of the packets flowing in the system.

3.3   Factors affecting performance

There are many factors that affect the performance of communication systems. Examples include spread spectrum, modulation, frequency hopping rate, packet size, offered load, error correction, transmission power, and network topology. In this list of examples, there are two types of factors: (1) factors affecting only the performance of the system where they are implemented, and (2) factors that affect the closed-loop interactions between the interferer and the victim's systems. The factors that belong to the first category include modulation scheme and error correction. All the others belong to the second category. In the context of this discussion, we are mostly interested in exploring factors belonging to the second category, since these factors alter the closed-loop behavior between the victim and the interferer systems. Factors belonging to the first category will be investigated in the context of solutions for coexistence.

Although performance results are invariably tied to the "victim's" performance, and, as discussed previously, every system is both a victim and an interferer, we consider two vantage points for every factor depending on whether it is implemented in the victim's system or in the interferer's system. Performance results are used to aid in the discussion and to illustrate the general trends observed. This approach is useful in answering two fundamental questions: (1) What features of the victim system affect its performance? (2) What features of the interferer system affect the performance of the victim system? Figure 3.5 presents an illustrative example of the approach considered.

The remainder of this section includes examples discussing the effects on performance of several parameters such as spectrum spreading, hop rate, offered load, packet size, transmission power, and network topology. In all examples presented, we assume a four-node topology including a victim communication system and an interferer communication system. The performance is measured at the victim's receiver, as illustrated in Figure 3.5. Keep in mind that the exact numerical results presented here are dependent on the specifics and the parameters of the scenario

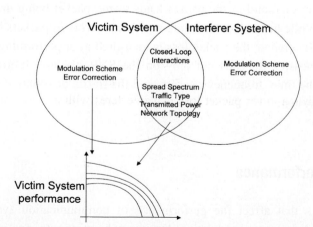

Fig. 3.5. Factors affecting the closed-loop interactions between victim and interferer systems.

considered. However, these results are still useful in highlighting trends and trade-offs in performance.

3.3.1   Spectrum spreading

As we saw in Chapter 2, there are two types of spectrum spreading techniques, namely direct sequence and frequency hopping. Although these two techniques will be revisited in the context of the solution space (Chapter 7), it is important to observe that the frequency hopping mode is generally less prone to interference than the spread spectrum mode. The probability of packet loss for frequency hopping systems is generally less than the one obtained for spread spectrum systems simply because by changing frequencies a device may be able to avoid others occupying the same band, while by occupying the same frequency band for a longer period of time a device may be hit more often, namely by a frequency hopper.

Now let us look at an example to illustrate this observation. Consider the packet loss of a victim system in Figure 3.6. A higher packet loss is obtained if the victim system has implemented direct sequence for spreading and if the interferer system has implemented frequency hopping. On the other hand, a victim system that has implemented frequency hopping and is being interfered with a direct sequence system has a lower packet loss. It is expected that this trend applies in most systems, although the relative position of each curve and the point of intersection between the two curves depicted in Figure 3.6 will most likely depend on the specifics of the technologies studied.

3.3.2   Hop rate

Having said that a victim system is less prone to interference if it implements frequency hopping, and a frequency hopping system can cause more interference on a victim system, next we explore the effects of the hopping rate of the victim and the interferer on the victim's performance. As expected, increasing the hopping rate of the victim system reduces the effects of interference, while it has the potential of causing more inteference on the other system. Figure 3.7 clearly illustrates this trend.

A frequency hopper has a lower packet loss if it interferes with a slower hopper. A faster hopper causes the packet loss to be substantially higher. This pattern should hold true for most systems, although the exact numerical results depend on the specifics and the parameters of the systems considered.

3.3.3   Traffic characteristics: offered load and packet size

Traffic patterns play a key role in a system's performance, and in the context of interference performance analysis this is no exception. Consider the traffic patterns

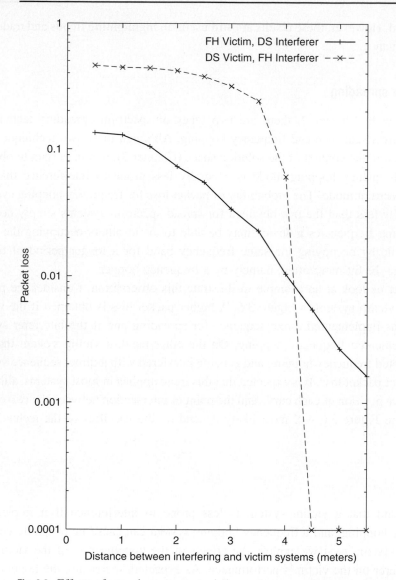

Fig. 3.6. Effects of spread spectrum on victim performance: Frequency hopping (FH) versus direct sequence (DS).

of two systems A and B as depicted in Figure 3.8. The transmission time indicates the interval during which a packet is being transmitted, while the idle time refers to the time when no packet is being transmitted. Observe that the relative position of the transmission time intervals between two systems directly affects interference and the resulting performance. Basically, the overlap between the transmission intervals of two different systems represents interference. This overlap, represented as a shaded area in Figure 3.8, causes the receiver of system A to receive incorrectly the packet destined to it because it can also "hear" the transmission of system B.

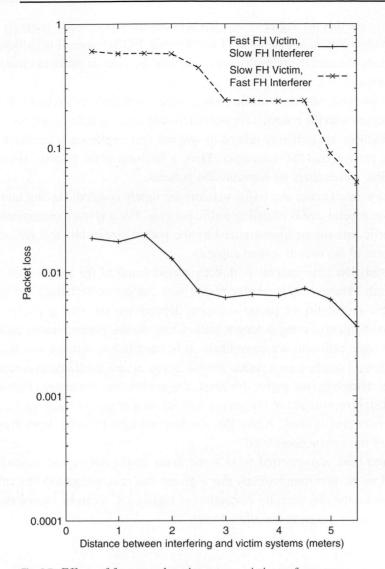

Fig. 3.7. Effects of frequency hopping rate on victim performance.

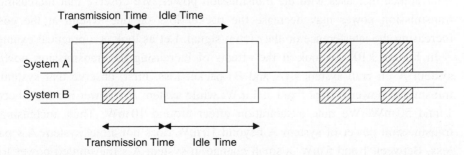

Fig. 3.8. Traffic patterns.

It is easy to see that the degree of overlap between the transmission intervals from different systems determines the level of interference. Furthermore, it is sufficient to alter the length of either the transmission or the idle intervals in order to change the level interference.

Now that we understand how traffic patterns affect interference, a related observation is concerned with the potential for interference to alter the traffic patterns. These mutual interactions are primarily related to systems that implement a feedback loop between the receiver and the transmitter. Thus, a collision at the receiver leads to a retransmission, which alters the transmission patterns.

In summary, interference and traffic patterns are tightly coupled. Having said that, let us explore general trends related to traffic patterns. For a given transmission data rate, the traffic patterns are characterized by the packet size in bits and the offered load in percent of the overall system capacity.

The transmission time interval is directly proportional to the packet size for a given data rate. Thus, using a shorter packet size can generally reduce the packet loss since the probability of packet collision depends on the time a packet is in transmission. Similarly, using a longer packet size makes systems more prone to interference since collisions are more likely. It is important to observe that the data rate has an impact on the time a packet spends in the air and the transmission length of a packet. Therefore, for higher bit rates, the packet loss decreases. Figure 3.9 gives an illustrative example of the packet loss for shorter packet sizes used for the victim and interferer systems. Again the absolute numerical results depend on the parameters of the system considered.

The offered load, also referred to in some cases as the duty cycle, is inversely proportional to the idle time interval. For a given data rate, the greater the offered load, the smaller the idle time. By inspection of Figure 3.8, it can be shown that the packet loss is proportional to the offered load.

3.3.4   Transmission power

Interference increases with the transmission power. We observe that increasing the transmission power may decrease the packet loss on one system at the cost of increasing the interference on the victim signal. Let us look at a practical example.

In Figure 3.10, we look at the effects of increasing the transmission power of system A on both system's (A and B) packet loss. First, observe that system B's transmitted power is kept fixed at 1 mW, while system A's power is varied between 1 and 50 mW. We note a saturation effect around 10 mW. Thus, increasing the transmission power of system A beyond 10 mW does not affect system A's packet loss. Between 1 and 5 mW, a small change in system A's transmitted power triples system B's packet loss.

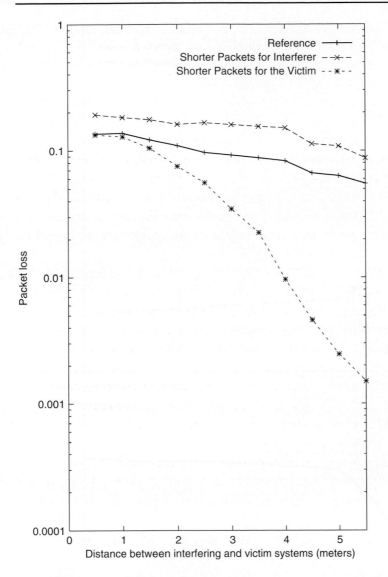

Fig. 3.9. Effects of packet size on victim performance.

Note that the relative positions of the packet loss curves are dependent on the parameters and the specifics of systems considered; however, the general trend that can be derived is as follows. As the transmitted power increases, the packet loss of the system under consideration may decrease, at the cost of increasing the interference level on the other system. Basically, we note that increasing the transmission power does not necessarily improve the performance. However, decreasing the transmission power is usually a "good neighbor" strategy that helps reduce the interference on other devices.

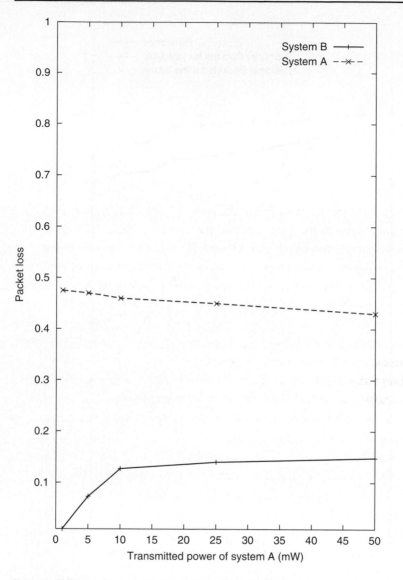

Fig. 3.10. Effects of transmitted power on performance.

3.3.5   Number of systems

Interference is proportional to the number of systems operating in the band. A system in this case refers to at least a pair of devices belonging to same network, i.e. a Bluetooth piconet, or a WLAN with two or more devices operating on a single center channel. Adding more Bluetooth piconets increases the WLAN packet loss. Similarly, increasing the number of WLAN systems (operating on different non-overlapping center frequencies) increases the interference on a Bluetooth piconet.

4 Interference modeling: open loop

Modeling interference is difficult since it requires considering simultaneously the interactions between the interferer and the victim system. This is known as closed-loop performance modeling and will be discussed in greater detail in Chapter 4. In this chapter, we discuss several approximations to evaluating interference at the victim's receiver while ignoring the interactions between the interferer and the victim system. This open-loop evaluation is generally a lot easier to model.

The first model we discuss deals with approximating the interference with white noise and deriving a probability of bit error at the receiver. This theoretical bit error calculation depends on the modulation, the energy of the signal transmitted, and the noise or interference levels. While not feasible in real implementations, most receiver designs include a probability of bit error calculation in the function of the signal to noise ratio. The waterfall shape of the bit error curve helps determine the optimal operating point for a particular design.

The second model is concerned with an n-state Markov model for characterizing the state of the wireless channel. This technique goes back to early work by Gilbert [41] that models the wireless channel using a two-state Markov chain where one state corresponds to a noisy channel and the other state corresponds to a noise free channel. We show how this model can be modified in order to model an interference limited and an interference free channel.

Finally in the last section of this chapter, we consider a probabilistic packet error model that relies on analysing the collision of multiple packets at the receiver including a mix of interferer and victim packets. The probability of packet collision is based on the timing and the number of simultaneous packet transmissions.

4.1   Theoretical BER estimation

Errors at a receiver system are due to a combination of thermal and electrical noise in addition to interference and a variety of channel impairments that distort the signal received and lead to retrieving inaccurately the signal that is transmitted.

In this section we discuss how bit errors are estimated at the receiver of a communication system given thermal noise, which is distributed according to a normal or Gaussian distribution. Observe that the power spectral density of thermal noise is $G_n(f) = N_0/2$ for all frequencies. A bit error rate curve is thus computed in terms of the signal to noise ratio and uniquely identifies the modulation and detection system used in the communication system receiver as illustrated in Figure 3.3. By approximating interference to noise in the signal to noise ratio, it is sometimes possible to use this waterfall shaped curve in order to get a first order approximation of the bit errors seen at the receiver for a given inteference level.

The bit error rate estimation for Gaussian noise follows. If we let $s_i(t)$ be the signal transmitted and $r_i(t)$ be the signal received, then we can write

$$r_i(t) = s_i(t) + n(t); \qquad \text{for} \quad i = 1, \ldots, M \tag{4.1}$$

where $n(t)$ is assumed to be the zero mean additive white Gaussian noise process.

In a binary system, where $i = 2$, the transmitted signal is simply given by

$$s_i(t) = \begin{cases} s_1(t) & \text{for a binary 1} \\ s_2(t) & \text{for a binary 0} \end{cases} \tag{4.2}$$

At the receiver system, sampling the received signal $r(t)$ means making a decision on whether s_1 or s_2 was transmitted, or, in other words, whether a binary one or zero was transmitted. We let z be the sampled signal over an interval T,

$$z(T) = a_i(T) + n_0(T); \quad i = 1, 2 \tag{4.3}$$

where $a_i(T)$ is the expected signal received and $n_0(T)$ is the noise component.

The probability density function of n_0 is given by

$$p(n_0) = \frac{1}{\sigma_0 \sqrt{2\pi}} \exp\left[-\frac{1}{2}\left(\frac{n_0}{\sigma}\right)^2\right] \tag{4.4}$$

where σ^2 is the noise variance. Thus, the conditional probabilities $p(z \mid s_1)$ and $p(z \mid s_2)$ can be written as follows:

$$p(z \mid s_1) = \frac{1}{\sigma_0 \sqrt{2\pi}} \exp\left[-\frac{1}{2}\left(\frac{z - a_1}{\sigma}\right)^2\right] \tag{4.5}$$

and

$$p(z \mid s_2) = \frac{1}{\sigma_0 \sqrt{2\pi}} \exp\left[-\frac{1}{2}\left(\frac{z - a_2}{\sigma}\right)^2\right] \tag{4.6}$$

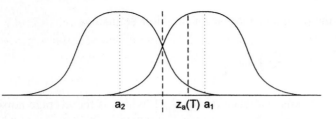

Fig. 4.1. Conditional probability density function.

These probabilities are illustrated in Figure 4.1 and represent the probability density of $z(T)$ given that s_1 was transmitted (rightmost curve) or s_2 was transmitted (leftmost curve).

A threshold γ is used in order to decide whether s_1 or s_2 was transmitted:

$$z(T) = \begin{cases} a_1 & \text{if } z(T) > \gamma \\ a_2 & \text{otherwise} \end{cases} \tag{4.7}$$

Typically γ is chosen in order to minimize the probablity of error. So if the likelihood $p(z \mid s_i)(i = 1, 2)$ are symmetrical then

$$\gamma = \frac{(a_1 + a_2)}{2} \tag{4.8}$$

where a_1 is the signal component of $z(T)$ when $s_1(t)$ is transmitted and a_2 is the signal component when $s_2(t)$ is transmitted.

There are two ways that an error can occur. An error occurs if a_2 is selected for a transmitted signal $s_1(t)$ or if a_1 is selected for a transmitted signal $s_2(t)$. The probability of bit error, P_B, is expressed as follows:

$$P_B = P(a_1 \mid s_2)P(s_2) + P(a_2 \mid s_1)P(s_1) \tag{4.9}$$

Since $P(s_2) = P(s_1) = 1/2$ and the probability density functions $P(a_1 \mid s_2)$ and $P(a_2 \mid s_1)$ are symmetrical,

$$P_B = \int_{\gamma}^{\text{inf}} p(z \mid s_2) \, dz \tag{4.10}$$

$$P_B = \int_{\gamma}^{\text{inf}} \frac{1}{\sigma_0 \sqrt{2\pi}} \exp\left[-\frac{1}{2} \left(\frac{z - a_2}{\sigma_0} \right)^2 \right] \tag{4.11}$$

where σ_0^2 is the variance of the noise. By letting $u = (z - a_2)/\sigma_0$, we can write

$$P_B = Q\left(\frac{a_1 - a_2}{2\sigma_0} \right) \tag{4.12}$$

where $Q(x)$ is the complementary error function defined as

$$Q(x) \approx \frac{1}{\sqrt{2\pi}} \int_x^{\inf} \exp\left(-\frac{u^2}{2}\right) du \qquad (4.13)$$

We can replace the noise variance σ^2 by $N_0/2$, which is the average noise power. The signal components a_1 and a_2 can also be replaced by the corresponding energy per bit transmitted, E_b. For a binary system, where $a_1 = \sqrt{E_b}$ and $a_2 = -\sqrt{E_b}$, P_B is given by

$$P_B = Q\left(\sqrt{\frac{2E_b}{N_0}}\right) \qquad (4.14)$$

Figure 4.2 plots Equation (4.14).

Thus, once the probability of bit error curve is derived for a system, it is easy to estimate the error rate at the receiver for a given E_b/N_0 level or signal to noise ratio (refer to Chapter 3 for correspondence). This curve is unique to every system since it depends on the modulation and the detection system used, and its theoretical derivation is similar to what was discussed in this section. It is beyond the scope of this book to cover the derivation of the probability of bit error for all communication systems. This discussion can be found in most communications textbooks at various level of detail; for example, see ref. [75].

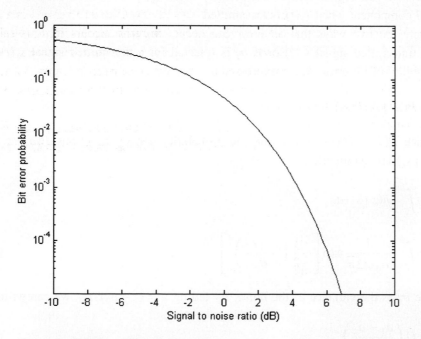

Fig. 4.2. Probability of bit error for a binary system.

As we stated at the beginning of this section, if the interference power were to be approximated with white Gaussian noise, then this probability of bit error curve can provide a rough approximation of the errors caused by interference. The error in approximating interference with noise can be significant since the power spectral density of interference is unlikely to be flat like Gaussian noise. Another problem with this approximation is that, unlike Gaussian noise, interference from multiple systems cannot be added at the receiver.

4.2   Modeling BER in wireless channels

In this section, we are concerned with developing an interference model that can be used as an input in evaluating the performance of a whole system, or studying specific functionality at higher layers, such as error correction, packet retransmissions, access control, and so on. In this case, the main objective is to focus on the impact of interference on the performance of the system. Therefore, BER is assumed as an input based either on experimental measurements or theoretical models.

Basically, the idea is to abstract the details of the communication channel and develop a parametrized model that accounts for errors due to wireless effects such as noise, fading, shadowing, and interference. The probability of error, P_e, is the main input parameter, although there may be other second order statistics that are used as well. The value for P_e is typically determined by analysing experimental measurement data using either a trace-based approach [59,60] or fitting techniques [54]. Additional experimental measurement and modeling techniques for the RF channel and the performance of wireless systems in noise and interference environments are found in the Telecommunications Industry Association reports [16,17].

There are two main flavors of models depending on whether or not error occurrence is time and state dependent. If errors are random (due to noise) and do not depend on time or state information, this model is known as the finite state memoryless channel. For binary systems, P_e is the probability that a 1 is incorrectly received as a 0. If the same probability of error, P_e, is used for characterizing errors in receiving a 0, then the binary channel is said to be symmetric [72].

On the other hand, a Markov model is generally used when the occurrence of errors is correlated in time or depends on state information. This model goes back to the early work by Gilbert [41], who used a two-state Markov chain where one state corresponds to a noisy channel and the other state to a noise free channel. This model was improved by Fritchman [39] and later by Wang and Moayeri [65] to characterize the envelope of a flat Rayleigh fading process.

The Gilbert–Elliot model [41] is probably the simplest model to consider for a wireless channel between two stations. This model is shown in Figure 4.3. It assumes

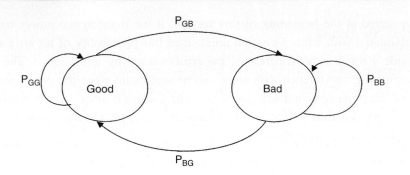

Fig. 4.3. Two-state error model.

an error state denoted by "Bad" and an error free state denoted by "Good". The probabilities associated with the "Good" and "Bad" states are P_G and P_B, respectively.

Next we define the transitional probability, P_{BG}, as the probability of being in "Bad" and transitioning to "Good", and P_{GG} as being in "Good" and staying in "Good", respectively. Similarly, we define P_{GB} as the transitional probability for being to state "Bad" and transitioning to state "Good" and P_{BB} as remaining in state "Bad". Therefore, we can define the transitional probability matrix as follows:

$$\begin{bmatrix} P_{GG} & P_{GB} \\ P_{BG} & P_{BB} \end{bmatrix}$$

The Markov model framework allows us to answer several questions about the system. In particular, how much time the system spends in each state, and what the probability is of being in a particular state. For discrete time Markov chain, state transitions occur after each channel symbol. If a packet model is used, then a transition is assumed for each packet. In this case, the time the system spends in each state is geometrically distributed. This time can be approximated by an exponential distribution for a large number of symbols or packets and for a relatively high probability of staying within a state.

Given that

$$P_B + P_G = 1 \tag{4.15}$$

we can compute the probability P_B to be given by

$$P_B = P_B P_{BB} + P_G P_{BG} \tag{4.16}$$

Replacing P_G by $1 - P_B$ we obtain

$$P_B = \frac{P_{BG}}{1 - P_{BB} + P_{BG}} \tag{4.17}$$

4.3   Packet error model

A third approach for modeling interference is concerned with associating a BER measurement with system timing and behavior information. A BER measurement is used as an input parameter in conjunction with the application traffic distribution, including packet size and interarrival time, in order to quantify the impact of interference on the system's performance. As a result, this approach provides an approximation model for the packets received in error at a particular node for a given BER and packet distribution model. The BER measurement used can be based on experimental measurements or vendor data sheets. It may also be useful to keep it as a parameter, in order to show the correlation of packet errors with respect to different BER values. There are two packet distribution models that need to be considered in this approach. First, we need to consider the packet distribution model used to characterize the behavior of the system under study. The system under study is basically the system that we are interested in obtaining a packet error model for. That is also the system being interfered with. Secondly, we need to consider the interfering system's packet distribution model. The objective of the method is to correlate these two packet distribution models and derive a probabilistic packet collision model that, for a given BER, yields packet errors. It is important to note, however, that the model does not consider any closed-loop interactions. In other words, it assumes that the resulting packet error does not affect the packet distribution model considered, although in most systems retransmissions implemented in layers 2 and above tend to alter the resulting packet distribution. These interactions can only be captured in simulation modeling, a topic that will be covered in greater detail in Chapter 5.

Before we describe the details of the derivation, let us go over some terminology. First, we consider two systems, namely the victim system and the interferer system. Packets sent and received by the victim system, A, are desired packets, while packets sent by the interferer system, B, are labeled interfering packets. The packet error model considers the probability of receiving a packet in error at the victim system assuming that it is being interfered with from the interferer system. A similar model can also be derived for the interferer system. Observe, however, that only one system is considered at a time in this model. This is why it is considered an open-loop model.

Next we explain step by step the modeling approach that can be used to derive a probability of packet error.

We consider a system A transmitter and receiver nodes as our reference and derive the probability that a packet containing errors (at least one error), $Pr(PE)$, is received at the victim system receiver node.

A collision occurs when both the desired and the interfering packets overlap in time and frequency. This collision is detected at the victim receiver in the form of SIR that depends on the power transmitted, the distance traveled, and the path loss

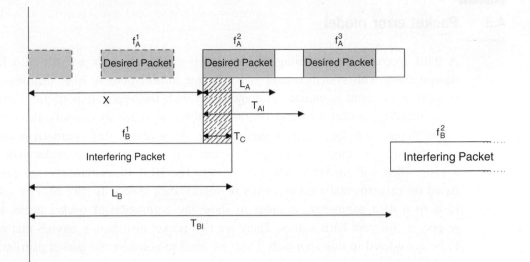

Fig. 4.4. Collisions at the victim's receiver.

model used. The SIR then translates into a BER according to the carrier modulation and the receiver implementation used.

Figure 4.4 illustrates the timing of the desired packets with respect to the interfering packets as seen at the victim's receiver. Let f_A and f_B be the frequencies used to transmit the desired and interfering packets, respectively. We denote by T_A and T_B the victim and the interfering packet transmission periods, respectively. In order to determine the position of the desired packet with respect to the interfering packet when both systems use the same frequency ($f_A = f_B$), we define a variable X that represents the time offset between a desired and an interfering packet. Let T_C represent the time interval when both interfering and desired packets overlap. We denote by T_{BI} the interval between two interfering packets including the packet transmission time T_B and a packet interarrival time. The packet interarrival time may include a backoff period, or account for several other variables. Similarly, we denote by T_{AI} the interval between two desired packet transmissions. We assume that X is a random variable that is uniformly distributed between zero and T_{BI}. Note that X is a continuous random variable; however, in this analysis it is quantified to the resolution of a bit duration:

$$X \sim U(0, T_{BI}) \tag{4.18}$$

Thus, the probability that a desired packet overlaps in time and frequency with an interfering packet depends on
- the position of the interfering packet with respect to the desired packet, i.e. X;
- the transmission frequencies, f_A and f_B of the victim and the interfering systems, respectively.

Note that X, f_A, and f_B are assumed to be independent and uniformly distributed random variables. The probability mass function of X is equal to $Pr_X(k) = 1/T_{BI}$, where $k = 1, 2, \ldots, T_{BI}$. Here, T_{BI} is expressed in bit durations or microseconds. Next we deal with the frequency overlap. Whether signals overlap in frequency depends on the bandwidth of f_A, and f_B, denoted by N_A and N_B, respectively, and on their relative position with respect to each other given the spectrum available bandwidth C. Let n be the number of frequencies of signal A affected by the transmission of signal B. Let δ_f be the frequency offset between the desired signal center frequency and the interfering signal center frequency, as depicted in Figure 4.5, where P_I and P_R refer to the interfering and received power, respectively. Now, $n = 2 \times \delta_f$, since δ_f is an absolute value and frequencies affected by interference can be located at $\pm \delta_f$. Note that n and SIR determine the BER received.

In other words, as signal A is closer to signal B in spatial and spectral distance, the BER obtained is higher. Similarly, as signal A moves away from the interfering signal center frequency and location, the BER curve drops sharply. This will be explained further when we look at an example in Section 4.3.1.

The probability that an interfering signal using f_B lands on the same frequency f_A depends on a discrete random variable f whose probability mass function is $Pr_f(j) = n/(C - N_A + 1)$ where j varies between 1 and $C - N_A + 1$. Expressing $Pr(PE)$ as a joint probability of frequency and packet overlap yields

$$Pr(PE) = \sum_{k=0}^{T_{BI}} Pr(PE \mid X = k; f = j) Pr_X(X = k) Pr_f(f = j) \qquad (4.19)$$

where $Pr(PE \mid X = k; f = j)$ depends on T_C and BER. Thus, we write

$$Pr(PE \mid X = k; f = j) = 1 - (1 - \text{BER})^{T_C} \qquad (4.20)$$

Therefore,

$$Pr(PE) = \left(\frac{n}{C - N_A + 1} \right) \left(\frac{1}{T_{BI}} \right) \sum_{k=0}^{T_{BI}} (1 - (1 - \text{BER})^{T_C}) \qquad (4.21)$$

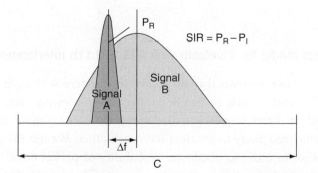

Fig. 4.5. Frequency offset.

The value of T_C depends on X, T_B, and T_A. We distinguish three cases.

(i) $T_A \leq T_B$ and $T_A \leq T_{BI} - T_B$

$$T_C = \begin{cases} T_A & \text{if } X < T_B - T_A \\ T_B - X & \text{if } T_B - T_A \leq X < T_B \\ 0 & \text{if } T_B < X \leq T_{BI} - T_A \\ X + T_A - T_{BI} & \text{if } T_{BI} - T_A < X \leq T_{BI} \end{cases} \tag{4.22}$$

(ii) $T_A \leq T_B$ and $T_A > T_{BI} - T_B$

$$T_C = \begin{cases} T_A & \text{if } X < T_B - T_A \\ T_B - X & \text{if } T_B - T_A \leq X < T_{BI} - T_A \\ T_B + T_A - T_{BI} & \text{if } T_{BI} - T_A \leq X < T_B \\ X + T_A - T_{BI} & \text{if } T_B \leq X \leq T_{BI} \end{cases} \tag{4.23}$$

(iii) $T_A > T_B$. We let $N(X)$ be the number of interfering packets that hit a desired packet:

$$N(X) = \begin{cases} \lceil \frac{T_A}{T_{BI}} \rceil & \text{if } X \leq T_{BI} \lceil \frac{T_A}{T_{BI}} \rceil - T_A \\ \lceil \frac{T_A}{T_{BI}} \rceil + 1 & \text{otherwise} \end{cases} \tag{4.24}$$

We also define T_i as the time collision of packet i:

$$T_i = \begin{cases} \max(T_B - X, 0) & \text{if } i = 1 \\ T_B & \text{if } i = 2, \ldots, N(X) - 1 \\ \min(X + T_A - (N(X) - 1) \times T_{BI}, T_B) & \text{if } i = N(X) \end{cases} \tag{4.25}$$

In this case T_C is basically the sum of the errors resulting from $N(X)$ colliding interfering packets:

$$T_C = \sum_{i=1}^{N(X)} T_i \tag{4.26}$$

4.3.1   Case study: packet error model for Bluetooth with IEEE 802.11b interference

As an example, we choose two systems that are known to interfere with each other, namely the Bluetooth system, which is based on frequency hopping, and IEEE 802.11b, which is based on direct sequence spread spectrum. Both systems use the 2.4 GHz band and therefore are likely to interfere with each other. We use the packet error model derived earlier in order to obtain the probability of packet error for the Bluetooth system. The analysis for this case study appeared in ref. [44]. The same

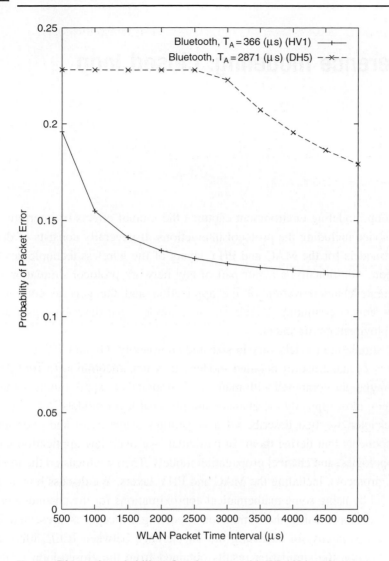

Fig. 4.6. Bluetooth probability of packet error with WLAN interference.

approach can be used for determining the packet error model for the IEEE 802.11b system.

Bluetooth has a frequency hopping span of 79 MHz, that is seventy-nine channels of 1 MHz width. Therefore, $C = 79$, $N_A = 1$ and $N_B = 22$; n is the number of Bluetooth frequencies affected by the presence of a WLAN interfering signal. Note that, unlike the signal bandwidth, which is fixed (22 MHz for WLAN DSSS), n depends on SIR for a given BER threshold. Using $n = 18$ and BER $= 0.24$, Figure 4.6 presents the probability of packet error for Bluetooth for varying WLAN packet sizes, T_B. The time interval between two WLAN packets, T_{BI} is set to twice $T_B \cdot n$ and BER can be computed from vendor fact sheets or experimental measurements.

5 Interference modeling: closed loop

A closed-loop modeling environment captures the mutual effects of interference on each end device including the protocol interactions. It generally consists of detailed simulation models for the MAC and PHY layers of the wireless technologies under consideration. Additionally, a major part of any network protocol simulation model is the accurate characterization of the application and the general configuration considered. This is commonly known as usage models that describe a user activity and the deployment environment.

Detailed simulation models vary in size and complexity. Given CPU and memory requirements to make use of detailed models, it is not uncommon to find detailed protocol simulations combined with mathematical models to approximate some parts of the system, most typically the channel and physical layer models.

In this chapter, we first describe what constitutes usage cases and overview the major components that define them. In particular, we overview application models, network topologies, and channel propagation models. Then we focus on the modeling of network protocols, including the MAC and PHY layers. We discuss how to speed up this model by using some mathematical approximations for the channel and PHY models. An example using the simulation modeling concepts is presented in the context of a case study for assessing the interference between IEEE 802.11b and Bluetooth. Finally, the simulation results obtained from the closed-loop model are compared with the open-loop model results described in Chapter 4.

5.1 Usage definition

The terms usage scenarios, use or usage models, and simulation scenarios are used somewhat interchangeably in the literature. They refer to model definitions that support the evaluation of a system's functional requirements. It is useful to distinguish between two broad categories of usage definitions, namely those leading to a qualitative evaluation of the technology requirements, and others that focus more on the quantitative aspect of the requirements and their evaluation. The first category is usually referred to as usage models, while the second category is known as simulation scenarios.

5.1.1   Usage models

Usage models typically describe the motivation, the functional use, and the deployment of a particular network technology or set of technologies. They identify the users of a technology, for example soldier, paramedic, healthcare provider, and the setting or the environment of the technology deployment, for example battlefield, emergency response, office, home, etc. They also describe the tasks that can be performed using the functionality provided. For example, a soldier uses a Wi-Fi connection in order to communicate with his battalion commander. A taxi driver uses a Bluetooth enabled hands-free headset device to carry a phone call. They typically answer the following questions.

 (i) Who is the user?
 (ii) What is the usage environment?
(iii) How and why is it useful?

5.1.2   Simulation scenarios

Simulation scenarios represent a finer level of description granularity with additional details. They are intended for the unambiguous definition of models that can be used in simulation modeling and performance evaluation. A simulation scenario is concerned with the choice of the application and the configuration and dimensioning of the network topology. It also includes details that can be derived from the usage model elements. The user and the task elements lead to the application characterization. On the other hand, the environment is linked to the topology and device positioning in addition to the channel propagation properties. We will now focus on the definition of simulation scenarios since we are concerned with a quantitative evaluation of network performance in the context of interference. Sections 5.2–5.4 are devoted to describing the application, topology, and channel models, all of which are part of the simulation scenarios.

5.2   Application models

Application models characterize a user's activity and the type of communication traffic that is generated. Since wireless devices are expected to carry traffic from as many communication services as possible, we anticipate a wide variety of applications ranging from voice, video, to data (for example FTP, email, web browsing, printing, and file sharing). A representative set of traffic models is defined to cover the range of applications envisioned. Rarely can a single model fit all usage scenarios; thus, there is a need to tailor traffic models to the specifics of each experiment's objectives.

There are three broad categories of traffic models, including (1) bulk data, (2) application profiles, and (3) traffic traces, and we provide a description for each category below.

5.2.1   Bulk data

This category is essentially an on–off packet generator source as depicted by Figure 5.1. This packet generator is characterized by an "on" state during which n packets of size l are generated. During the "off" state no packet is generated. The time intervals t_{on} and t_{off}, the source spends in the "on" and "off" state, respectively, generally represent means of a statistical distribution, for example exponential, paretto, or other. Similarly, the number and the size of the packets generated can also satisfy some distributions.

The main objective in using a bulk data model is to capture the non-continuous nature of packet transmission and not necessarily the exact behavior of higher layer protocols such as TCP/IP, HTTP, FTP. This may be interesting in the context of controlled experiments where traffic parameters such as packet type, size, and interarrival time need to be isolated and their effects on interference investigated. Therefore, rather than implementing the entire stack above the MAC layer, the on–off model is used with parameters to characterize the traffic burstiness and the user's activity as seen at layer 2, such as packet interarrival and packet size. These two parameters are in turn described by suitable distribution functions.

This approach to ignore the details of higher layer protocol implementation, including signaling messages such as TCP connection establishment and other network discovery commands that may be sent over the physical channel, is common in the evaluation of layer 2 protocols. These are modeled by the statistical attributes of the traffic.

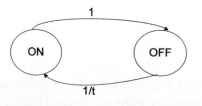

Fig. 5.1. On–off traffic source.

A simple traffic source often used in the evaluation of MAC protocols consists of generating packets of fixed length l, according to an exponential interarrival time, t_s, that is proportional to the overall medium capacity,

$$t_s = \frac{(l/d)}{\lambda} \tag{5.1}$$

where d is the data rate in bits/s used to transmit l bits and λ is the offered load as a percentage of the total link capacity.

Let us look at an example on how to compute the packet interarrival time for the WLAN protocol. Assuming an IP packet of 1500 bytes, and a MAC packet payload of 12 000 bits, the mean packet interarrival time as a function of the offered load is given by

$$t_W = \left(\frac{192}{1\,000\,000} + \frac{12\,000 + 224}{d} \right) \lambda \tag{5.2}$$

where 224 is the MAC layer header, 192 is the PHY layer header (sent at 1 Mbit/s). The payload is transmitted at d, which is either 1 or 11 Mbits/s.

It is also worth to note here that these simple on–off models can in fact account for layering overheads. For example, in order to transport an IP packet over a MAC layer, the application traffic can be segmented into IP packets and those packets in turn can be segmented and packaged into MAC packets. IP packets exceeding the specific MAC payload size are segmented into smaller chunks before being transmitted over the medium.

5.2.2   Application profiles

Profiles are used to model the behavior of different application types, for example FTP, HTTP, video, print, email, and voice. The first step in devising profiles is to identify parameters that characterize the application behavior. The second step is to configure the parameters and choose realistic values that characterize different operating conditions.

Profile parameters are identified according to the protocol implementation and the configuration considered. Although profile parameters are directly associated with the protocol characteristics they are supposed to model, they depend on the specifics and the level of detail of each implementation. Regardless of the level of detail chosen, the expectation is that at least part, if not all, of the transport and signaling protocols in the TCP/IP stack are captured.

As an illustrative example, let us consider the profile parameters of an FTP application; these mainly comprise two operations in an FTP application related to file transfer, namely (m)put to upload a file (or multiple files) onto a server and (m)get to download a file (or multiple files) from a server. In addition, there are other

Table 5.1. *Example of parameter profile*

Parameters	Distribution	Value
FTP		
Traffic directionality	server to client	get command
Inter-request time (s)	exponential	1
File size (bytes)	constant	2M
HTTP		
Page interarrival time (s)	exponential	5
Number of objects per page	constant	2
Object 1 size (bytes)	constant	10K
Object 2 size (bytes)	uniform	(200K, 600K)

operations related to opening and closing the connection, for example directory listing operations. Parameters related to file transfer operations typically include the inter-request time and the file size. The inter-request time is the interval between two FTP commands, and the file size represents the size of the file requested in bytes.

Similarly, a profile for an HTTP application can be devised. HTTP profile parameters are used to characterize a web page request such as the page interarrival time, the number of objects on each page, and their size in bytes.

While identifying application profile parameters generally represents a first step, choosing appropriate parameter values represents the next phase in the profile modeling step. The parameter values depend on the usage scenario that is envisaged given a particular network technology and topology.

An example of a parameter profile set is given in Table 5.1. These parameters capture the traffic burstiness and patterns of the traffic traces for FTP and HTTP obtained from realistic WLAN connections.

5.2.3   Traffic traces

A third type of traffic sources can be derived from observing and capturing packet traces from an operational network. This requires an operational network set-up and tools that record the packets exchanged between two nodes. The minimum configuration set-up consists of two terminals (desktops, laptops, or other devices) equipped with network interface cards (for example WLAN, Bluetooth). There are several commercially and publicly available tools to do the packet logging. Commonly used tools for logging TCP/IP layer packets are WinDump running on Windows and TCPDump operating on Unix/Linux.[†]

[†] Information about TCPDump can be found on the following web site: http://www.tcpdump.org/

Table 5.2. *FTP capture statistics*

Parameters	Average value
Bluetooth	
Traffic directionality	put command
Inter-request time (s)	5
File size (bytes)	2M
WLAN	
Traffic directionality	put command
Inter-request time (s)	1
File size (bytes)	2M

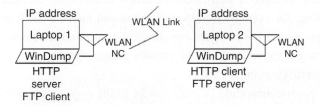

Fig. 5.2. Traffic trace capture and configuration set-up.

The computers are configured to operate in ad hoc mode so that a communication link is set up between them, as shown in Figure 5.2.

Application traffic such as HTTP or FTP are recorded using TCPDump or WinDump and later analysed in order to obtain statistics concerning the packet size observed. These statistics constitute the traffic traces and include information about each packet being transmitted by a node, such as application type, port number, packet size, time it is generated, and source and destination address. All of this information catpured about the packets transmitted can then be used to model the application. It can also serve to derive statistics for the packet size distributions and the interarrival times. Alternatively, it can help in the generation of packets according to the packet timing and size recorded. In fact, many network simulation tools have application program interface (API) to integrate input directly from traffic traces.

Table 5.2 summarizes the statistics collected on an FTP trace using two different network interface cards, namely a Bluetooth card and a WLAN card using the set-up in Figure 5.2. Both traces were obtained with 2 Mbyte file transfers. For the Bluetooth FTP trace, 5 seconds were used between two file transfers, while for the WLAN FTP trace, 1–2 seconds were used between two file transfers. The average time to transfer a file is around 0.5 seconds.

5.3   Network topology

In this section, we discuss what constitutes a network topology and the details required for using and defining a network topology for the purpose of configuring a simulation scenario.

A network topology ranges in size and complexity depending on the scenario studied. The simplest network topology possible consists of two nodes. One node represents a transmitter while the other node represents a receiver. The transmitter and receiver may be also referred to as source and sink nodes, respectively. For the quantification of interference, the simplest topology consists of four nodes: two transmitter–receiver pairs as depicted in Figure 5.3. The number of nodes in a topology can be as little as three if non-communication devices are considered in the interference modeling, for example microwave ovens can interfere with wireless communications. Simple network topologies constitute a controlled environment for investigating the effects and the interactions of various parameters that are varied as part of the simulation configuration set-up.

To study interference, communicating devices can be of the same or different type. Thus, the device's description includes protocol and technology specific parameters, for example MAC and PHY layer specifications used, transmitted power level, receiver's sensitivity, packet encapsulation type. Basically, all parameter settings that differ from the expected defaults are usually identified and their value explicitly specified.

Next comes the placement of the devices on a two-dimensional grid. Each node is then associated with x and y coordinates, as shown in Figure 5.3.

The relative position of the devices and the transmitted power determines a coverage area that is depicted by the two circles in Figure 5.3.

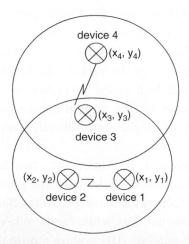

Fig. 5.3. Four-node network topology.

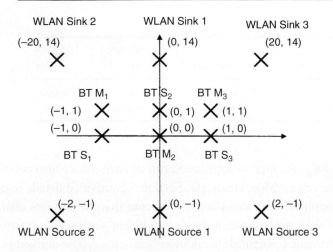

Fig. 5.4. Multi-WLANs and Bluetooth piconets.

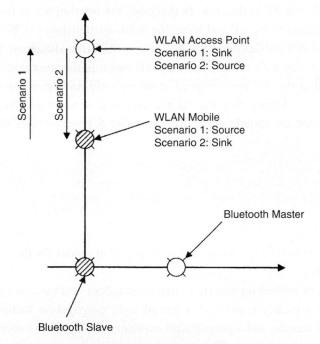

Fig. 5.5. Traffic flow directionality.

From a relatively simple topology of four nodes, other topologies can be constructed, including multiple transmitter/receiver pairs as shown in Figure 5.4. It consists of three WLAN systems (source–sink pairs) and three Bluetooth piconets, each with one master and one slave device. It represents a more stringent interference environment with multi-Bluetooth piconets and multi-WLANs.

Table 5.3. *Traffic directionality scenarios*

Scenario	Desired signal	Interferer signal	WLAN AP	WLAN mobile
1	Bluetooth	WLAN	sink	source
2	WLAN	Bluetooth	source	sink

As part of the topology, the type of application run on each device also needs to be specified, for example voice, video, email, etc. Section 5.2 provided details regarding the modeling of the application space. There is no doubt that the accurate characterization of the application running on each device significantly affects the interference and the resulting performance obtained. In addition, scenarios can be derived in order to limit the flow directionality in order to control further the interference measured on a receiver device. For example, in Figure 5.5 we let the mobile be the generator of IEEE 802.11 data, while the AP is the sink. In this case, the interference is from the mobile sending data packets to the AP and receiving acknowledgments (ACKs) from it. Since most of the WLAN traffic is originating close to the Bluetooth piconet, both the master and the slave may suffer from serious interference. In the second scenario, the traffic is generated at the AP and received at the WLAN mobile. Because the data packets are generally longer then the ACKs, this is a more critical scenario for the WLAN than when the mobile is the source. Table 5.3 summarizes the two scenarios.

5.4 Channel model

The channel model consists of a geometry based propagation model for the signals, as well as a noise model.

Channel models can be broken up into two categories: indoor and outdoor channel models. These models typically consist of a line-of-sight propagation factor (free space) for the first few meters, and a propagation exponent for higher distances.

A general form for the path loss in dB is given by

$$L_\mathrm{p} = \alpha + \beta \log_{10}(d) \tag{5.3}$$

where d is the distance between the transmitter and the receiver in meters and α and β are parameters determined by the model. Values for α and β are determined from experimental measurements of specific radio systems. Table 5.4 gives typical path loss exponents based on experimental measurements.

Table 5.4. *Typical path loss exponents*

Environment	β
Outdoors, free space	2
Outdoors, shadowed urban area	2.75–5
In building, line of sight	1.6–1.8
In building, obstructed	4 to 6

Assuming unit gain for the transmitter and receiver antennas and ignoring additional losses, the received power in dBm is given by

$$P_{\mathrm{R}} = P_{\mathrm{T}} - L_{\mathrm{p}} \tag{5.4}$$

where P_{T} is the transmitted power, also in dBm. Equation (5.4) is used for calculating the power received at a given point due to either interferer or victim transmitters, since this equation does not depend on the modulation method.

The main parameter that drives the PHY layer performance is the signal to interference ratio between the desired signal and the interfering signal. This ratio is given in dB by

$$SIR = P_{\mathrm{R}} - P_{\mathrm{I}} \tag{5.5}$$

where P_{I} is the interference power at the receiver.

To complete the channel model, noise is added to the received samples, according to a specified SNR. In decibels, the signal to noise ratio is defined by $SNR = P_{\mathrm{R}} - S_{\mathrm{R}}$, where P_{R} is the received signal power and S_{R} is the receiver's sensitivity in dBm; this latter value is dependent on the receiver model and so is an input parameter. A number of noise models including additive white Gaussian noise (AWGN), multipath fading, can be used to model the noise at the receivers.

5.5   Protocol layer modeling

In this section we describe the MAC and PHY layer modeling approaches since they require distinctly different simulation techniques. We also discuss how to interface both models in order to create a simulation of the entire system.

5.5.1   MAC layer modeling

A number of event driven simulation packages available in the public or commercial domain can be used to model MAC protocols. In event driven simulations time is used

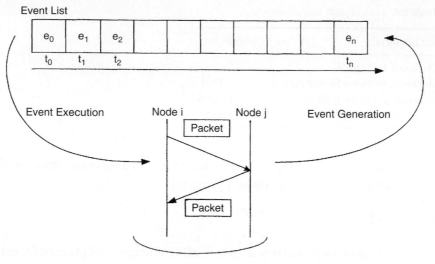

Protocol behavioral description

Fig. 5.6. Event driven simulations.

as a global variable in order to keep track of events and execute them according to a pre-determined order. Time is advanced to the time of the next scheduled event in the event queue. It can be advanced by an arbitrary amount rather than by a fixed increment as in time driven simulations, as shown in Figure 5.6.

Since the main MAC protocol functionality consists in processing packets, events are generally triggered following a packet arrival and the need for the protocol to perform some processing functions, either to manipulate the packet or to update internal state variables. Other MAC protocol functions include the execution of transmission rules, segmentation and reassembly, backoff and contention resolution, acknowledgment, retransmission, and scheduling, as covered in Chapter 2.

MAC protocols are generally modeled as a communication state machine, where each state represents a different phase in the communication protocol. Figure 5.7 shows a state machine for a generic MAC protocol. States constitute different phases in the protocol execution. The so-called wait states are those awaiting the arrival of an event or set of events. For example, in the idle state, the protocol waits for either a higher layer or physical layer packet arrival. In the case of a higher layer packet arrival the next state must perform channel sensing before transmitting the packet. Both the channel sensing and the transmit state are transitional and are associated with a specific MAC operation as opposed to a wait state. If the medium is detected as busy, the MAC goes into a backoff state awaiting the expiration of a timer before attempting a transmission. After transmitting a packet and in the case of an acknowledged service, the protocol waits for the receipt of an ACK packet.

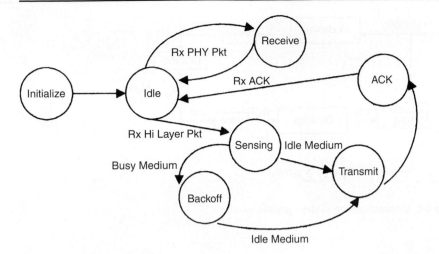

Fig. 5.7. Generic MAC state machine.

5.5.2   PHY layer modeling

Digital signal processing level modeling is required to simulate the wireless device physical layer including the transmitter, receiver, and the channel propagation properties.

Unlike in the modeling of MAC protocols, where the processing of discrete events is the most important function, the notion of time is essential in the modeling of communication physical devices such as signal modulation, filtering, and detection and their functionality.

The basic idea behind the simulation of a communication system consists of generating sampled values of all the input waveforms, and then processing these samples through the different functional blocks discussed in Chapter 2 such as modulation, filtering, and demodulation.

This process is referred to as a Monte Carlo simulation. From Figure 5.8, the input and noise signals, $A(K)$ and $N(K)$, respectively, are assumed to be random processes. The objective is to find the statistical properties of $Y(t)$ or some function of $g(Y(t))$. The expected value of $E(Y(t))$ is computed as follows:

$$P_e = \frac{1}{N} \sum_{k=1}^{N} g(Y(k)) \tag{5.6}$$

The overall process works as follows. For a given transmitter, samples are generated at a certain sampling rate that is chosen to provide several samples per symbol for the technology considered. The received samples from both the desired transmitter and the noise or interferer(s) are added together at the receiver. The receiver looks at the received symbol and decides which was the most likely transmitted one based on the specifics of the decoder procedure implemented.

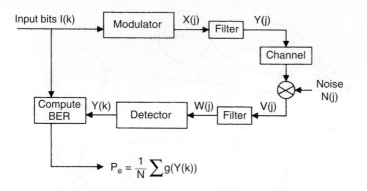

Fig. 5.8. Communication system generic model.

5.5.3   Packet and signal processing simulation models interface

Putting it all together, the packet level simulation models are interfaced to the signal processing models in order to provide a complete evaluation platform. The step-by-step simulation process works as follows.

Traffic is generated by sources located above the MAC layer. The message is then passed to the MAC layer, where it undergoes encapsulation and obeys the MAC transmission rules. At the end of each packet transmission, the MAC layer generates a data structure that contains all the information required to process the packet. This structure includes a list of all the interfering packets with their respective duration, timing offset, frequency, and transmitted power. The topology of the scenario is also included. This data structure is then passed to the physical layer along with a stream of bits representing the packet being transmitted. This constitutes the input parameters depicted in Figure 5.9. The physical layer returns the output parameters in the bit stream after placing the errors resulting from the interference.

Consider the transmitter–channel–receiver chain of processes in the physical layer. For a given packet, the transmitter creates a set of signal samples that are corrupted by the channel and are input to the receiver; interference may be present for all or

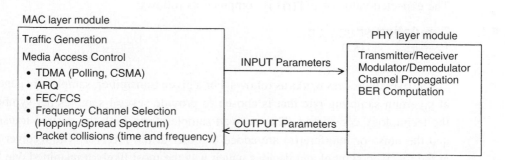

Fig. 5.9. MAC/PHY simulation interface.

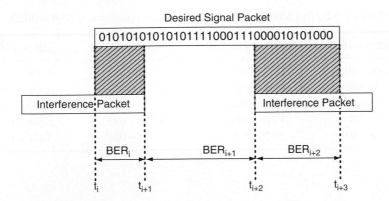

Desired Signal Packet

Fig. 5.10. Period of stationarity.

only specific periods of stationarity. For interference to occur, packets must overlap in both time and frequency. In a system with many interfering systems, there may be interference from more than one packet at any given time. We define a period of stationarity (POS) as the time during which the interference is constant.

During a POS, as shown in Figure 5.10, where there is one or more interferers, the number and location of bit errors in the desired packet depend on a number of factors: (1) the signal to interference ratio (SIR) and the signal to noise ratio at the receiver; (2) the type of modulation used by the transmitter and the interferer; and (3) the channel model. For this reason, it is essential to use models that accurately characterize the channel and the PHY layer. Just because two packets overlap in time and frequency does not necessary lead to bit errors and the consequent packet loss. While one can use (semi-) analytic models instead, the use of detailed signal processing-based models better allows one to handle multiple simultaneous interferers.

The physical layer returns the bit stream after placing the errors resulting from the interference. At this point, the MAC layer applies the error correction algorithm corresponding to the packet encapsulation used before deciding whether to accept or drop a packet.

5.6 Simulation speed-up

In order to speed up the simulation process, each transmitter–channel–receiver process can be replaced with a table-based approach combined with a binary symmetric channel. BER tables for different values of SIR and for different frequency offsets Δf can be derived. To create tables such as Table 5.5, the curves are sampled for fixed steps in both SNR and SIR.

For a segment of a packet where the interference is stationary, the SNR and SIR are computed using the transmitters' powers, the topology, and the path loss model.

Table 5.5. *BER for an 802.11b* 1 Mbit/s *receiver and Bluetooth interference;* $SNR = 30$ dB

Δf	SIR $= -16$ dB	SIR $= -15.5$ dB	SIR $= -15$ dB	SIR $= -14.5$ dB
0	3.69e − 001	3.56e − 001	3.74e − 001	3.20e − 001
1	4.11e − 001	4.24e − 001	4.17e − 001	3.99e − 001
2	3.82e − 001	3.85e − 001	3.61e − 001	3.48e − 001
3	4.05e − 001	4.22e − 001	3.57e − 001	3.87e − 001
4	3.74e − 001	3.62e − 001	3.38e − 001	3.44e − 001

Thus, using the calculated SIR and the given frequency offset of the intended signal with respect to the interference signal, the average BER can be extracted by a simple table lookup operation. Errors are then generated for each bit of the packet segment using the binary symmetric channel with crossover probability equal to the average BER of the segment. The SNR in these tables is assumed to be very high (greater than 30 dB), which is the case for interference-limited environments. However, the software can check this assumption by comparing the SIR to this value.

Using tabulated BER values, as opposed to running the detailed signal processing receiver and channel simulation models in real time, is expected to give a speed-up factor. However, in order to verify that the speed-up gained does not jeopardize the accuracy of the results, simulation results for both the MAC and PHY models should be compared and validated against analytical given different traffic scenarios.

5.7   Case study: evaluating IEEE 802.11 and Bluetooth interference

Having defined all the pieces of the simulation modeling approach, and developed detailed simulation models for the IEEE 802.11 and Bluetooth interference case, let us use them in order to evaluate the resulting network performance. We use the network topology presented in Figure 5.3, where the WLAN mobile is located at $(0, d)$, and d is varied from 0.5 to 5 meters. The sources of the WLAN data are the WLAN mobile in scenario 1 and the WLAN AP in scenario 2, as shown in Table 5.3. Traffic for WLAN is bulk data, where the offered load, λ, is set to 50%. Similarly for Bluetooth, bulk data are used for data traffic with $\lambda = 50\%$. The simulation parameters are summarized in Table 5.6.

Figure 5.11(a) gives the packet loss for scenario 1, where the Bluetooth piconet is closer to the WLAN transmitter. The effects of the WLAN 11 Mbit/s interference on Bluetooth leads to slightly higher packet loss (20%) for Bluetooth data compared

Table 5.6. *Simulation parameters*

Bluetooth parameters	
Bulk data	$\lambda = 50\%$, DH5 packet size = 2870 bits
Voice	$\lambda = 100\%$, HV1 packet size = 366 bits
Transmitted power	1 mW
WLAN parameters	
Bulk data	$\lambda = 50\%$, packet size = 12 224 bits
Transmitted power	25 mW

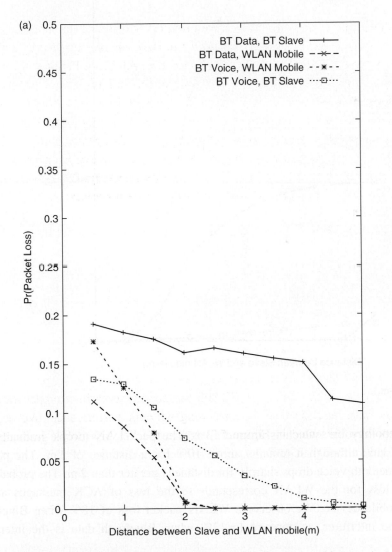

Fig. 5.11. Bluetooth and 802.11 DSSS 11 Mbit/s interference. (a) Scenario 1: Bluetooth piconet close to the WLAN transmitter. (b) Scenario 2: Bluetooth piconet close to the WLAN receiver.

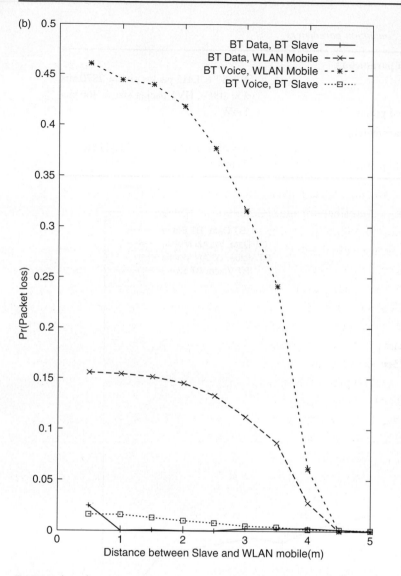

Fig. 5.11. (cont.).

with Bluetooth voice, which is around 13%. The packet loss drops gradually for Bluetooth data, although it remains above 10% for a distance of 5 m. The packet loss for Bluetooth voice drops sharply for distances greater than 2 m. The probability of packet loss for the WLAN corresponds to the loss of ACK messages at the WLAN mobile device. We observe a WLAN packet loss of 18% when Bluetooth voice is the interferer, as opposed to 12% when Bluetooth data is the interferer signal.

Figure 5.11(b) gives the packet loss for scenario 2. Note that the packet loss when the WLAN receiver is close to a Bluetooth voice connection (45%) is triple that when it is close to a Bluetooth data connection (15%). However, the WLAN 11 Mbit/s packet loss remains greater than 25% until a distance of 4 m. The packet loss for Bluetooth is negligible in this case since the WLAN source is far from the Bluetooth piconet (15 m) and does not affect the receiver.

5.8   Comparing the simulation results with the analysis

Now that we have discussed a closed-loop simulation model that can accurately characterize the effects of interference, it is worthwhile comparing it with the open-loop model described in Chapter 4.

Recall that we obtained an expression for $Pr(PE)$. Our objective is to compare the numerical results obtained from the analysis with those obtained from the simulation models of the systems corresponding to the same parameters.

In particular, we would like to verify that the probability of packet error analysis can provide a close approximation to the system's packet error and packet loss measures. Consider the four-node topology consisting of two Bluetooth devices and two IEEE 802.11b devices. Let us compare the probability of packet error on the Bluetooth slave device computed according to Equation (4.21) and obtained using the detailed MAC and PHY simulation models described in Section 5.7. We use the topology and configuration parameters shown in Figure 5.3 and Table 5.6, respectively.

For the Bluetooth signal we assume a pair of devices: a master and a slave device located at (1, 0) and (0, 0) m, respectively. For the WLAN signal, we use two 802.11 direct sequence devices transmitting at 11 Mbits/s. We assume unidirectional traffic; a WLAN source transmits packets to a WLAN sink that returns ACK messages to the source. The WLAN source and sink devices are located at $(0, d)$ and $(0, 15)$ m, respectively, where d is varied from 0.5 to 5 m.

A couple of observations are in order to justify the correlation between the simulation parameters and the parameters used in the analysis. First, we compute the SIR from the distances and transmitted powers as follows. Given that the WLAN source is at a distance $d_{\mathrm{I}} = d$ m from the Bluetooth slave, while the Bluetooth master is at a distance $d_{\mathrm{M}} = 1$ m, and assuming that the WLAN source and the Bluetooth master device transmit at 25 mW and 1 mW, respectively, the SIR in dB at the slave is given by $10 \log \frac{1}{25} + 20 \log(d_{\mathrm{I}}/d_{\mathrm{M}})$ according to the logarithmic path loss model given in Equation (5.3) with $\alpha = 0$ and $\beta = 2$. Table 5.7 summarizes the SIR for different values of d.

The choice of the BER value and n corresponding to this SIR is then derived from the PHY results of the Bluetooth receiver. An example of PHY layer performance results is given in Figure 5.12, which depicts the BER for a Bluetooth receiver with WLAN interference for varying SIR and frequency offsets. The reader is referred to ref. [63] for additional performance results on this Bluetooth receiver. The frequency offset is

Table 5.7. *SIR computation*

d (m)	SIR (dB)	BER	n
0.5	−20	3.68×10^{-1}	18
1	−14	3.41×10^{-1}	16
2	−7.95	1.37×10^{-1}	16
3	−4.43	7.6×10^{-2}	16
4	−1.93	4.2×10^{-2}	14
5	0	3.1×10^{-2}	10

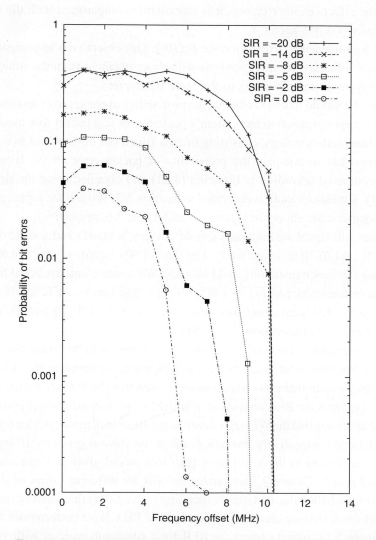

Fig. 5.12. Impact of WLAN interference on Bluetooth. Probability of bit error plotted against frequency offset.

defined as the distance in megahertz between the desired signal (Bluetooth) and the interfering signal center frequency. We note that as the Bluetooth signal is closer to the WLAN signal center frequency (small frequency offset), the BER is higher. Similarly, as the Bluetooth signal moves away from the WLAN signal center frequency, the BER curves drop sharply. Therefore, the range of frequency offsets with a high BER determines n, which is in this case the number of Bluetooth channels that may be affected by the presence of WLAN interference. Here we summarize the details of the procedure used for selecting BER and n. For a given SIR, we set a BER threshold value, BER^T, and look up the corresponding frequency offset, ΔF, from Figure 5.12. For example, for $SIR = -20 \, dB$ and $BER^T = 10^{-1}$, ΔF is equal to $9 \, MHz$. We note that

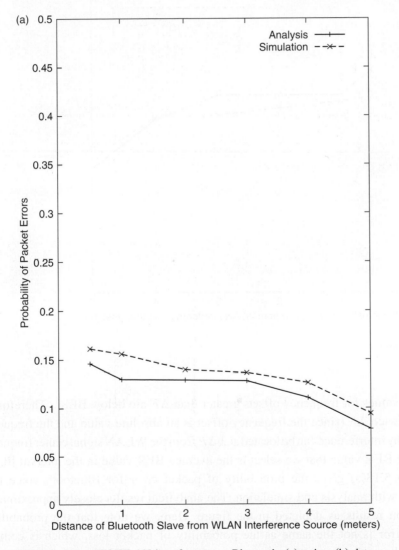

Fig. 5.13. Impact of WLAN interference on Bluetooth: (a) voice; (b) data.

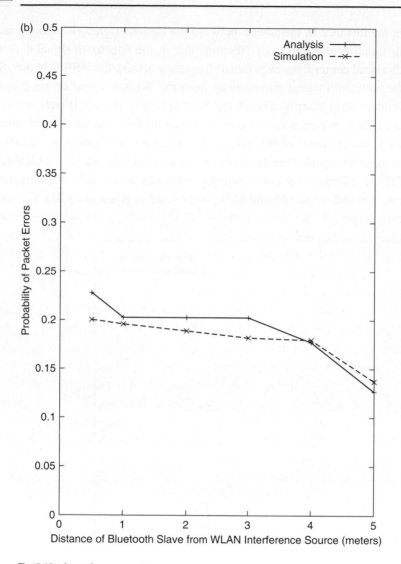

Fig. 5.13. (cont.).

all BER values for frequency offsets greater than ΔF are below BER^T. Therefore, we set n to twice ΔF (since the frequency offset is an absolute value and the frequencies affected by interference can be located at $\pm\Delta F$ from the WLAN signal center frequency). Also, the BER value that we select is the average BER value in the interval $[0, \Delta F]$.

Figure 5.13(a) gives the probability of packet error for Bluetooth voice traffic obtained with analysis and simulation. The analytical results closely approximate the simulation results as depicted in the figure. Here, we note that the probability of packet error is not the same as the probability of packet loss, which is computed after applying the appropriate error correction mechanisms. Therefore we expect the probability of packet error to represent an upper bound on the probability of packet loss, since this latter measure depends on the error correction code used.

Figure 5.13(b) depicts the analysis and simulation results for the Bluetooth data traffic. We note that the probability of packet error for data traffic is higher than that for voice traffic mainly because data packets are longer than voice packets (2871 bits for data packets as opposed to 366 bits for voice packets). Note the difference ($\sim$5%) between the analytical and the simulation results for distances up to 3 m. We attribute this to the closed-loop interference effect present in the simulation but not modeled in the analysis. Basically, the WLAN source causes interference on the Bluetooth devices; as a result, there are more retransmissions within the Bluetooth piconet. This in turn causes ACK losses at the mobile and WLAN packet retransmissions. This mutual interaction causes changes in the traffic distribution for both WLAN and Bluetooth systems and makes the analytical results more pessimistic (up by 5%). On the other hand, for distances above 4 m, where the interference is low, the analytical and simulation results are in good agreement.

Finally, we conclude that the probability of packet error analysis, in the tractable case where mutual interference effects can be ignored (as in the case of Bluetooth voice traffic or in the case of low interference), can provide a close approximation to the packet error obtained from the simulation models. The packet error model derived provides an upper bound on the probability of packet loss.

Ignoring the effects of mutual interference makes the analysis more pessimistic since it does not consider how interference causes changes to the traffic patterns and the overall MAC behavior. In the case studied, there is about a 5% difference between the analytical approximation and the simulation results.

There may be scenarios where this technique can provide an accurate approximation; for example when the source of interference is mainly a transmitter and receives very few or only ACK packets. In addition, this method may be useful in providing a "back of the envelope" approximation in cases where bit error rates are available from vendor fact sheets.

6 Channel estimation and selection

An accurate assessment of channel conditions represents a first step in any interference mitigation strategy. This channel assessment is generally performed at the receiver and used by the transmitter in order to make an informed decision about the channel state. In some cases, the channel assessment is also performed at the transmitter's side. This is common in most carrier sense multiple access systems, where devices have to "listen" to the medium before transmission. In this case, the transmitter and the receiver devices are assumed to be closely located and therefore have the same channel conditions. However, it is not uncommon for the receiver side to be experiencing different channel conditions from the transmitter. This situation is also known as the hidden node problem and conversely the exposed node problem in carrier sense multiple access sytems. Therefore, to optimize communication it is critical that each transmitter and receiver pair maintains the state of the channel as seen by the receiver.

Given this need to assess the channel conditions per transmitter and receiver pair, there are two basic channel estimation strategies. Channel estimation can be based on either explicit or implicit methods. Explicit methods include bit error rate calculation, packet loss, or frame error rate measurements performed on each receiver. The measurements are then conveyed to the transmitter device at regular time intervals. Alternatively, implicit methods do not require the transmitter and receiver devices to exchange information about the state of the channel. This information is derived by the transmitter based on the communication exchange, for example a negative ACK or an ACK timeout, and the need to retransmit the information. Note that either channel estimation method enables the transmitter device to avoid data transmission on a "bad" channel.

In this chapter, we first give an overview of the different types of measurements that can be used to assess the channel conditions. We then describe how to adapt these measurements dynamically in real time. Finally we discuss a three step procedure for the implementation of channel estimation in real systems.

6.1   Measurements

In this section, we discuss what performance measurements can be used to quantify the received signal quality. They are classified into two categories: measurements obtained from the PHY layer, and others from the higher layer (MAC layer and above).

6.1.1   Physical layer measurements

The most significant measurements characterizing the signal quality at the PHY layer include the measured power at the receiver. However, others are also available in order to account for the channel propagation conditions and the receiver's design.

   (i) Received power level (RPL) in dBm. This measurement represents the RF signal power as observed at the receiver's antenna. It includes noise and interference power levels.
  (ii) Noise and interference level (NIL) in dBm. This is typically an intermediate measurement in order to assess the noise and interference in the channel. It is performed when the node is idle; in other words, when it is neither transmitting nor receiving frames.
 (iii) Received signal to noise indicator (RSNI) in dbm. This represents a ratio of the received power level at the antenna to the noise and interference levels measured when the node is idle. It is defined as $RSNI = RPL - NIL$.
  (iv) Signal to interference ratio (SIR) in dB. This is a theoretical measure for the ratio of the power level of the signal at the receiver divided by the interference level at the receiver.
   (v) Transmit power level (TPL) in dBm. Unlike RPL, which is a power level measured at the receiver, this power level is made available to the receiver. This is a value that is encoded in the transmitted frame.
  (vi) Bit error rate (BER). This represents a theoretical figure based on the modulation and coding schemes used and the SIR. This value is either tabulated or obtained from vendor data facts sheets. It is computed at the output of the demodulator, but prior to performing frame error correction or checking if available.
 (vii) Frame error rate (FER). This represents the rate of frames in error after performing error correction and checking. It is defined as the ratio of frames received in error over the total number of frames received.

6.1.2   Higher layer measurements

These measurements are available at all layers above the PHY layer, but are generally associated with the MAC layer. They include a number of statistics that are kept at the transmitter and the receiver nodes.

(i) Number of packets transmitted (PTX). This is simply a counter that tracks the number of packets transmitted at the transmitter.

(ii) Number of packets received (PRX). This is a counter that tracks the number of packets at the receiver. It may include packets received in error and later discarded following an error check process.

(iii) Queue length. This represents the number of packets awaiting transmission at the MAC layer. A large queue length signifies that the system is operating beyond its capacity either due to a high offered load or bad channel conditions requiring retransmissions.

(iv) Number of packet retransmissions (PRTX). This is the number of times a packet received in error has to be retransmitted. This measurement is only possible when there is an ARQ mechanism at the MAC layer. In the absence of an ARQ mechanism, the transmitter has no way of telling if a packet was not successfully received at the other end, and therefore does not know to repeat its transmission. This measurement concerns a transmitted packet and not an acknowledgment packet.

(v) Packet loss (PLoss). This measurement represents the ratio of packets received in error to the total number of packets received. It is often used as an indicator of noise or interference in the channel.

(vi) Number of negative ACKs or ACK loss (NACK/ACKLoss). If an ARQ mechanism is in place at the MAC layer, this measurement implicitly tracks how many packets are received in error at the destination node. This is a useful measurement for the transmitter in order to derive the packets lost at the receiver without having to exchange this information explicitly.

(vii) Delay in seconds. This value measures the time it takes a packet to be transmitted and received at the receiver. It includes the transmission and propagation time.

(viii) Throughput in bit/s. This represents a ratio of the total number of packets received sucessfully over the time it took to receive the packet at the other end.

(ix) Jitter. This represents the packet interarrival delay variation. For real time applications such as voice and video, this is an important measure since packets are expected at fixed time intervals at the destination. Delayed packets that arrive "too late" are usually discarded.

Table 6.1 summarizes the measurements defined in this section.

6.2 Adaptive measurements

The effectiveness of the measurements discussed in Section 6.1 relies heavily on their accuracy in tracking the conditions prevailing in the channel including assessing the

Table 6.1. *Measurements summary*

Parameter	Definition	Node
PHY layer		
RPL (dBm)	power level at the antenna	receiver
NIL (dBm)	noise and interference level	receiver
RSNI (dBm)	signal to noise level	receiver
SIR (dBm)	signal to noise ratio	receiver
TPL (dBm)	transmit power level	transmitter
BER	bit error rate	receiver
FER	frame error rate	receiver
Higher layer		
PTX	number of packets transmitted	transmitter
PRX	number of packets received	receiver
QL	queue length	transmitter
PRTX	number of retransmitted packets	transmitter
PLoss	ratio of packets lost to packets received	receiver
NACK	number of negative acknowledgment packets	transmitter
ACKLoss	ratio of acknowledgment packets lost to acknowledgment packets received	transmitter
Delay (s)	end to end packet transmission and propagation delay	receiver
Throughput (bit/s)	ratio of bits to the time it takes to transmit them	receiver
Jitter	packet interarrival delay	receiver

presence of interference. Two key elements determine the effectiveness of measurements, namely how they evolve over time, and how they can be combined together in order to provide a better assessment of the channel conditions. These two issues will be discussed in the following two sections.

6.2.1   Time average

Although real time measurements are extremely desirable for timely control, reacting too quickly to an instantaneous measurement may lead to overall system instabilities and wide swings in performance. Observe that making use of these measurements comes at a significant overhead, which, depending on the solution sought, may require exchanging messages between the transmitter and the receiver, retransmitting a collided packet, delaying a packet transmission, or selecting a different transmission channel all together. Therefore, in order to keep the communication overhead down, it is critical not to react based on an instantaneous measurement but rather to use an average of observations collected over time. However, there is a trade-off between the time it takes to perform the measurements and the response time. Ideally a short observation time leads to a quick response time. Thus the system response time depends on the choice of the observation time. Having established the benefits of computing averages, we now turn to discussing several methods aimed at tracking statistics over time. They range in complexity from simple averages, to weighted averages and histograms.

(i) Simple average. This is the mean of a number of values collected over time. This average is computed by specifying a time window over which all values observed are averaged, or specifying a number of values to add. Let n be the number of values observed for a measurement v. Then, the average value, v_a, is computed as follows:

$$v_a = \frac{\sum_{i=1}^n v_i}{n} \tag{6.1}$$

(ii) Weighted average. As its name indicates, a weighted average is computed by giving different weights to the values observed. Although multiple weights can be considered, generally a weight is used to represent the previous value of the measurement observed and constitutes a form of memory. Let w be the weight factor such that $0 \leq w \leq 1$:

$$v_i = w * v_{i-1} + (1 - w) * v_i \tag{6.2}$$

where v_i is the value at time i and v_{i-1} is the value at time $i - 1$. Thus, w represents a built-in memory of sorts. As w approaches unity, the observed value is slowly combined to the running average. On the other hand a value of w close to zero represents an instantaneous measurement.

(iii) Histogram. If additional details are sought in order to track the measurement variations over time, histograms can be built. Histograms are useful in order to estimate better the variations of a measure over time. Thus measurements are categorized into a number of bins representing different values. As measurements are observed, they are added to the appropriate bin defined. In the case of discrete measurement, for example the number of retransmissions, a counter corresponding to the appropriate bin is simply incremented. For decimal values, bin values are bound by two thresholds. For example, in Figure 6.1 a histogram is constructed in order to track bit error measurements. Each bin is defined as all values included between two bit error thresholds.

6.2.2   Combining measurements

Another key element to using measurements effectively is the need to combine multiple observations in order to derive more accurate information about the channel. When attempting to identify interference it is important to distinguish between different types of environments and conditions that may all lead to a high packet

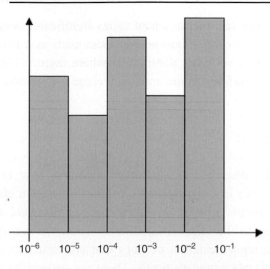

Fig. 6.1. BER measurement histogram.

loss. Therefore, using a single measure, for example packet loss, is not sufficient to assess the presence of interference. As an example, let us consider the following conditions all exhibiting a high packet loss: (1) a noise-limited environment, (2) an overloaded medium, (3) an out-of-range environment, and (4) an interference prone environment. Since the remedy sought for each of the conditions listed above may be different, it is important to identify correctly which case we are dealing with. Next, we discuss each of these conditions and provide clues in order to identify them and adequately deal with them. We assume that in all four cases packet loss is significant.

(i) A noise-limited environment is generally assessed if the noise level at the receiver is significant when the receiver is idle and at the same time the signal power level is high. The observed bit error rate does not vary over time; in other words, the instantaneous value for the bit error rate or the frame error rate is comparable to a weighted average or a time distribution for this measure.

(ii) A medium overload situation is when the queue length at the MAC layer transmitter becomes significantly large and the number of packets received exceeds the medium capacity. In this case a packet loss is due to a large number of packets colliding. Observe that packet loss occurs in contention-based systems. On the other hand, in reservation-based systems, packets are dropped at the transmitter before attempting to access the medium.

(iii) An out-of-range environment is identified when the power level of the signal at the receiver is weak and the noise level is high. In addition, the number of packets received at the destination is below the medium capacity.

(iv) An interference condition is assessed when the power level of the signal at the receiver is high, the number of received packets does not exceed capacity, and the instantaneous bit error rate measurement varies significantly over time. In addition, looking at a bit error distribution over time, such as a histogram, reveals a multimodal (at least bimodal) distribution where there is clearly an average for periods of non-interference and another average for the interference period.

6.2.3 Thresholding techniques

While Sections 6.2.1 and 6.2.2 described how to track measures over time and how to combine multiple measures in order to derive useful information about the channel conditions, we now discuss how to use these measures in real time in order to adapt the transmission to the channel conditions. Thus, the use of these measures is predicated on the implementation of flags that are raised as soon as measures reach or exceed acceptable boundary levels. These pre-defined levels, also known as thresholds, consist of pre-computed values that are set for acceptable network performance. They may include a bit error rate, a packet loss, a throughput, a queue size, etc. There are three methods one can think of in order to generate triggers.

  (i) Number of consecutive errors. In its simplest form, this thresholding technique tracks a counter and generates a trigger as soon as the counter reaches a pre-defined level. Most likely the counter is for a number of consecutive packets received in error or the number of consecutive packet retransmissions. This counter is reset as soon as the trigger is generated or a new condition occurs that negates the previous condition. For example, if the consecutive number of packets received in error is tracked and the threshold is set to four, then if three packets are received in error and a fourth packet is successfully received, the counter is reset and no trigger is generated. On the other hand, if a fourth packet is received in error, the counter is reset after a trigger is generated.

 (ii) Threshold level. This is a pre-defined value that determines an acceptable level of performance. If the measure tracked exceeds or falls below this level a trigger is generated. For example, if a queue length threshold is set to K, as soon as the queue length exceeds K a trigger is generated. The measure tracked could be an instantaneous value or an average.

(iii) Multithreshold levels. In order to ensure smooth system operation and to avoid oscillations, multithreshold levels are usually set; at least two levels are defined. For example, in the case of a bit error threshold, if the measured average bit error rate falls below a threshold, BER_1, then the channel is assumed to be error free. If, on the other hand, the bit error rate is greater than BER_2, the channel is assumed to be prone to interference. If the measured bit error rate is somewhere in between BER_1 and BER_2, no action is taken. This form of hysteresis is generally effective in combatting oscillations.

6.3   Implementation issues

Having dealt with the types of measures and how they can be combined to derive information or generate triggers, it is time to consider system implementation issues. Channel estimation consists of the following three step procedure.

(i) Channel classification. In this step, the channel is scanned in order to determine whether interference is present.

(ii) Feedback. This represents step 2. Since the channel conditions are usually different at the transmitter and the receiver nodes, information about the channel is exchanged between both nodes. The receiver informs the transmitter about the conditions prevailing at the receiver's end, unless the transmitter can derive this information.

(iii) Mitigation. In this third step the transmitter stops using the channel prone to interference. This is the interference mitigation step per se.

These three steps are repeated over and over again in order to adapt the interference mitigation procedures to varying channel conditions. Important considerations include how often the estimation is performed and for how long. Tyically, estimation is performed after a system is powered up, and following an idle period. An important question is, should channel classification be on-going while data is being transmitted? If the answer to this question is yes, then how often should channel classification be performed?

These questions are addressed in the following subsections in the context of a discussion of channel classification and feedback procedures. The interference mitigation step is left until Chapter 7, where it is presented as part of a comprehensive discussion of various available coexistence solutions.

6.3.1   Channel classification

Channel classification is based on measurements conducted on each frequency used in the band (or center frequency in the case of spread spectrum systems), in order to determine the presence of interference or noise. Using a simple thresholding method similar to the one in the example below, the channel is classified into two groups: "good" and "bad" channels. In this example, our discussion focuses on packet loss measurements; however, most other measurements described previously can also be used. In a nutshell, packet-loss-based channel estimation works as follows. Each receiver maintains a *frequency status table* (FST), where a percentage of packets dropped (due to errors), PLoss, is associated to each frequency, f, as shown in Figure 6.2. Frequencies are classified "good" or "bad" depending on whether their packet loss rate is below or above a threshold value, respectively.

Status	Frequency offset	Pkt Loss
good	0	10^{-3}
bad	1	0.75
bad	2	1
bad	3	0.89
	...	
good	76	10^{-4}
good	77	10^{-3}
good	78	10^{-3}

Fig. 6.2. Frequency status table.

In Figure 6.2 the threshold value is equal to 0.5. Each receiver node has its own FST maintained locally. In addition to its own FST, the transmitter has a copy of the receiver's FST.

For frequency hopping systems, such as Bluetooth, a simple rule such as the number of times each frequency is visited, N_f, can be used in order to derive the status of a frequency. This has the advantage of keeping the system more dynamic in that it may respond more rapidly to changes in the environment. Other systems that may use the same center channels, such as IEEE 802.11b, can rely on time averages.

6.3.2   Channel estimation feedback

At regular time intervals, each receiver updates its FST copy, which is kept at the transmitter, using a status update message that can be defined as part of the MAC layer protocol message exchange.

This constitutes the so-called explicit estimation, where explicit messages are exchanged between the receiver and the transmitter which contain information about the state of the channel at the receiver. However, in most cases where a handshake mechanism is available, the transmitter can derive information about the receiver's channel conditions based on negative acknowledgment or the lack of acknowledgment. Thus, the status of the channel at the receiver can be derived implicitly and can be kept locally. This is the so-called implicit channel estimation, where a receiver does not need to update the FST copy kept at the transmitter. Implicit estimation has two main advantages, the most obvious being the elimination of the communication overhead between the transmitter and the receiver. The second advantage is that it considerably reduces the response time. As soon as a frequency is determined to be "bad", it can be classified as such right away and potentially avoided in the next packet transmission. Thus, the time to react to an interference condition can be significantly reduced.

6.3.3   Channel estimation frequency

An interesting question we would like to answer is how to adapt the channel estimation procedure to changes in the environment. In particular, we would like to determine (1) the time it takes to perform the estimation and derive the FSTs, and (2) how often to perform the estimation and update the FSTs. Parts of this discussion appeared in ref. [42].

First, we define two phases in the channel estimation procedure. During the *estimation window*, EW, packets are sent on all frequencies regardless of their classification. Note that if no data traffic is available for transmission, probe and response type packets could be exchanged between the transmitter and the receiver in order to scan the channel and collect measurements. This message exchange is designed in most implementations to keep the connection alive and check the status of the receiver. It comes at the expense of causing more interference on other systems. The estimation window takes place at the beginning of every *estimation interval*, EI, and is followed by an online phase where only "good" channels are selected for the data transmission, as depicted in Figure 6.3.

We observe that during both estimation and online phases, PLoss is measured and continuously updated. Next, we give a lower bound on EW and describe how to adjust EI based on the environment's dynamics.

Estimation window

The time to perform the channel estimation depends on the transmission rate and the frequency hopping rate (in the case of a frequency hopping system), since the methods used to perform the classification depend on packet loss measurements per frequency visited. In the following we discuss a lower bound calculation for the Bluetooth frequency hopping system. First, we assume a hop rate of 1600 hops/s given single-slot packets. For each receiver the hopping rate is $1600/2 = 800$ hops/s since nodes receive on every other "frequency" or "hop" in the sequence. Next, we consider the Bluetooth frequency hopping algorithm. In a window of thirty-two frequencies, every frequency is selected once, then the window is advanced by sixteen frequencies, and the process is repeated. Therefore, it takes five windows of thirty-two frequencies in order to visit each of the seventy-nine frequencies twice. In other words, 160 hops visit each frequency twice. The time to visit each frequency four times per receiver

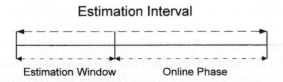

Estimation Interval

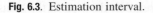

Estimation Window   Online Phase

Fig. 6.3. Estimation interval.

is given by $160/800 \times 2 = 0.4$ seconds $= 400$ ms. In fact, 400 ms constitutes a lower bound assuming a full load and single-slot packets.

In order to avoid having to fix EW, or compute it manually, we use simple techniques to adjust the window dynamically based on the number of times, N_f, each frequency in the band is visited. For example, if N_f is equal to 2, then each receiving frequency in the band is visited at least twice, i.e. the estimation phase ends only when the last frequency in the band has been used twice for each device in the piconet.

Estimation interval

How often we update the channel estimation depends on the application and the dynamics of the scenario used. Let us look at an adaptive procedure to adjust the EI, the interval between two consecutive estimation windows.

First, we let δ be the percentage of frequencies that change classification status (from "good" to "bad" or vice versa) during the previous estimation phase. More formally, let $S(f, t)$ be the status of frequency f at time t:

$$S(f, t) = 1 \quad \text{if } f \text{ is "good"}$$

$$S(f, t) = 0 \quad \text{otherwise} \tag{6.3}$$

Using the exclusive bit or operation between $S(f, t)$ and $S(f, t+1)$ represents the change of status of frequency f from time t to $t+1$. A change of status leads to a logic "1" while a no change yields a logic "0". Summing over all frequencies and dividing by the number of frequencies available, which is N in this case, is then equal to δ, as follows:

$$\delta = \frac{1}{N} \sum_{f}^{N} (S(f, t) \oplus S(f, t+1)) \tag{6.4}$$

We can then define a procedure for adapting EI. Initially, EI is set to $\text{EI}_{\min}$. Then it is updated every interval, k, according to the rationale that if a change were to happen it is likely to happen again in the near future and therefore EI is set to $\text{EI}_{\min}$. Otherwise, the window is doubled:

$$\text{EI}_{k+1} = \max(2 \times \text{EI}_k, \text{EI}_{\max}) \quad \text{if } \delta \leq 0.1$$

$$\text{EI}_{k+1} = \text{EI}_{\min} \quad \text{otherwise} \tag{6.5}$$

Now let us consider the implementation of the channel estimation procedure in the context of the chosen feedback mechanism . Recall that two feedback mechanisms can be used. One is based on an explicit message exchange between the transmitter and the receiver and the other is based on implicit information derived at the transmitter.

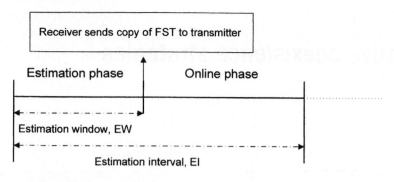

Fig. 6.4. Explicit estimation.

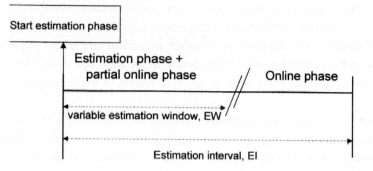

Fig. 6.5. Implicit estimation.

In the case of explicit feedback, the local FSTs can be updated every time a packet is received, while the copy of the receiver FST kept at the transmitter is only updated at the end of each estimation window, or at the beginning of the estimation interval, as shown in Figure 6.4.

On the other hand, in the case of implicit estimation, the boundary between the estimation window and the online phase is blurred. While the channel estimation procedure is still performed every EI, the size of EW no longer has to be pre-determined, as illustrated in Figure 6.5.

Although in both cases the estimation window is scheduled every EI, the online phase starts abruptly at the end of the estimation window in the explicit case, and ramps up gradually in the implicit case.

Thus, the implicit method can be faster in adapting to rapid changes in the interference environment. Also, the two phases become somewhat complementary. In the online phase, estimation is mainly used to determine what frequencies the transmitter should avoid using, while the estimation window phase is used to bring "good" frequencies back into use.

7 Effective coexistence strategies

The main goal of this chapter is to describe proven techniques employed to mitigate interference. We focus on dynamic and system level mechanisms that are able to adapt to the interference environment. Interference suppression techniques including coding, modulation, and filtering, in addition to others related to physical layer technologies such as CDMA and OFDM, abound in the open literature. The interested reader is referred to other texts such as ref. [76] for an in-depth treatment of these technologies.

The system level coexistence techniques that we are concerned with can be classified into two broad categories. The first category of solutions consists of some form of sharing, making use of either temporal or spectral sharing, and, in some instances, a joint time and frequency domain technique. The second category of solutions is about adaptation and the opportunity to choose either the radio or the network that is best suited to the environment. This class of solutions includes handovers and the ability to roam across different networks.

Sharing the medium is synonymous with multiple access techniques. In Chapter 2, we briefly overviewed three types of multiple access techniques, namely TDMA, FDMA, and CDMA. Since the emphasis is on system level solutions, we consider TDMA and FDMA. Although CDMA is considered to be an effective interference suppression technique, it is specific to coding functionality at the physical layer and therefore it is not pertinent to a system level solution. Furthermore, CDMA requires synchronization and control messages to be exchanged between the transmitter and the receiver, which has limited applicability in the context of wireless networks operating in the unlicensed bands.

A few observations on the approach we follow are in order before we proceed with the discussion of select coexistence techniques. Like all techniques developed to solve a problem, the techniques described in this chapter are developed in close relation to the technology and the system operation they are supposed to improve. In other words, to be effective coexistence solutions are tailored to the specific interference scenarios of interest in addition to the functionality of the victim and interferer systems involved. While cookbook recipes help produce the desired results, a discussion of specific solutions without outlining the general components and the methodology used may be of limited benefit outside the scope of the solution itself. Therefore, we try to address what most solutions have in common and look for generic components that have to be considered in the design of actual solutions. Illustrative examples are provided wherever applicable in order to demonstrate the use of the design principles and provide the reader with the tools needed in order to develop different solutions.

The remainder of this chapter is organized as follows. First, we describe how to detect the presence of interference in the environment. This represents a key component in all the techniques described in this chapter. Next we discuss time and frequency division multiple access techniques followed by handover mechanisms. For each class of solutions we present examples and performance results for select algorithms and scenarios of interest.

7.1   Knowledge of interference patterns

Knowledge of the interference patterns represents the most critical component of any coexistence solution. Basically, the effectiveness of time sharing the medium between the victim and the interferer system depends on how well the interference signal patterns are known to the victim system.

In a distributed environment, and in the absence of a central system where a node that knows everything manages others, the assumption is that there is no direct communication between nodes. Each system needs to know of the presence of other systems that are not of the same type. From an interference perspective, we observe that each system is both an interferer and a victim system. Therefore, each victim needs to obtain an estimate for the interferer's traffic. Chapter 6 provides the reader with a reasonably detailed discussion of measurements that can be used to estimate the channel or detect the presence of interference in the channel.

The development of solutions has to be tuned to whether knowledge of interference patterns is available and, if it is available, how accurate it is to the actual signal. Optimizing the time sharing often requires a complete and accurate knowledge of the interference traffic patterns. If only an approximation is available, the time sharing will tend to be rough, or it will underutilize the medium.

For example, in frequency hopping systems, such as the Bluetooth radio link technology, the presence of an IEEE 802.11b system leads to a higher packet loss on select channels, mainly those occupied by the IEEE 802.11b system. Therefore, instead of knowing the actual packet transmission times of the IEEE 802.11b system operating nearby, it is easier for the Bluetooth system to assume that these channels are occupied and therefore should no longer be used. A similar argument can be made for direct sequence spread spectrum technologies. For example, a low rate IEEE 802.15.4 device can detect the presence of an IEEE 802.11b system and tag all frequencies used by this IEEE 802.11b system as no longer available.

Without a direct communication between the devices, it is almost impossible to estimate the traffic patterns, much less to predict future transmission patterns for interference.

Another idea is to be able to differentiate between different types of systems and noise present in the channel. What this boils down to is how well the victim node can detect the presence of other systems in its vicinity, and most significantly know how to differentiate between random noise, other systems of the same type, and the presence of interference. Identifying these three different cases is critical to the successful implementation of a solution. Solutions that overcome random noise, or share the medium among contending systems of the same type, are often ineffective to mitigate interference. This concept will be revisited in greater detail in Chapter 8.

7.2   Time division multiple access

Time sharing the medium constitutes an effective interference mitigation technique. In fact, time sharing as part of the time division multiple access (TDMA) class of MAC layers is one way of sharing the medium among many users as described in Chapter 2.

At the heart of developing a time-sharing-based solution for interference mitigation are three main building blocks: (1) knowledge of the interference patterns, (2) fairness in sharing the medium, (3) QOS support.

Once the presence of interference is known, a device should delay its transmission until the medium is clear. In other words, the main concept behind time sharing is to let the interferer finish its transmission before a transmission is attempted. Thus, in a system some devices will be experiencing interference and therefore will not be able to use the medium at certain periods of time. In the case where there is no central reservation, for example in the case of a contention-based system, there is not much more to do except delay the transmission and contend for the medium once the interference is gone. However, in a reservation-based system, this missed transmission opportunity due to interference can be given to another device that is perhaps not experiencing the same type of interference. Therefore, the problem becomes one of how to skip the transmission turn of some users experiencing periods of interference while maintaining fairness and supporting the quality of service requirements.

Next we discuss how to achieve both of these objectives in a scheduling system. Observe that the concepts of fairness and quality of service support described here apply to any reservation-based system.

7.2.1 Fairness

An important criterion in time sharing is fairness. When resources are shared among many users, special attention is given to distributing the resources fairly among all users. A simple way to express fairness is to divide up the resources equally among all users. However, because not all users require the same amount and type of resources, then the fairness criterion is adjusted to take the diversity of the user requirements into account.

The term max–min fairness is introduced in Bertsekas and Gallager [69] in the context of flow control. It refers to giving the maximum bandwidth allocation to sessions with minimum requirements. These sessions are constrained in the sense that they will not be able to use any extra bandwidth. Therefore, once the allocation for the most constrained session is conducted, the bandwidth is distributed to other sessions that can consume extra bandwidth. Where there are several levels of constraints, allocation proceeds starting with the minimum constrained until all the bandwidth is allocated.

A formal definition of max–min fairness given in ref. [69] is as follows. Let r_i be the allocation given to a session i. Let B be the total allocation of bandwidth on a link with capacity C:

$$B = \sum_{i=1}^{n} r_i \tag{7.1}$$

Now, r_i is a feasible allocation if

$$r_i \geq 0 \quad \text{for all} \quad i \tag{7.2}$$

and

$$B \leq C \tag{7.3}$$

Note that r_i is max–min fair if it is feasible and it cannot be increased without decreasing the allocation of another session j crossing the same bottleneck and whose allocation $r_j < r_i$.

There are many algorithms that achieve max–min fairness, for example a leaky bucket. In its simplest form, a round robin algorithm can achieve max–min fairness if all sessions share the bandwidth available at the bottleneck link equally. In Section 7.2.3, we explore a credit-based mechanism for achieving max–min fairness.

Now that we have discussed the meaning of max–min fairness, it may be beneficial to describe how to measure it. Again, different measures may be considered for the so-called fairness index. In its simplest form, a fairness index measures how an actual allocation differs from an ideal or an optimum allocation. We choose a fairness index for each session, f_i, and a global fairness index, GFI, for the entire system. Formally, we define f_i as follows:

$$f_i = \frac{M_i}{E_i} \tag{7.4}$$

where M_i is the measured value for session i's allocation, and E_i is the expected optimal allocation; M_i and E_i can represent a number of packets, bits, or even a bit rate. On the other hand, it is often interesting to compute a fairness index for the entire system.

$$\text{GFI} = \frac{(\sum_i^n f_i)^2}{n \sum_i^n f_i^2} \tag{7.5}$$

GFI [33] is, by definition, bound between zero and unity and measures the fairness of the entire system based on the allocation of each session according to its requirements.

7.2.2   QOS support considerations

In the max–min fairness scenario described in Section 7.2.1, we made the simplifying assumption that all sessions are bandwidth hungry and are only constrained by having to share one bottleneck link. This may be different in some cases since some sessions may have different minimum requirements, and these requirements may be higher than those provided by the max–min fairness allocation. Therefore, the max–min fairness strategy can be revisited in order to consider the quality of service constraints of each session.

We mentioned in the Section 7.2.1 that a leaky bucket algorithm can be used to achieve a max–min fairness allocation. We now give an example of such an algorithm.

The main idea is to use a credit system in order to control the bandwidth allocated to each device and ensure that no device gets more than its fair share of the available bandwidth. Thus, each device i is given a credit c_i, and devices with a positive credit counter $(c_i > 0)$ are allowed to send data. Credits can be computed according to the negotiated rate or required rate for each session. Credits can represent a packet transmission, a transmission opportunity, or a number of bits. Credits are derived so that they have a direct relation with the QOS parameters considered, for example peak bandwidth, end-to-end delay, and jitter. For example, in order to compute credits in terms of bits for a session i requiring a rate of p bit/s, p credits should be allocated to session *i* every second. Additional examples on how to compute credits are given in the case study.

Once credits are allocated, they are decremented every time a packet or x number of bits are transmitted. The final part of a time sharing strategy is to allocate a servicing priority to users, so that instead of using the medium all at once, they can take turns using the medium. With a credit-based system it is easy to derive a servicing scheme that gives users with more credits a higher priority for transmission.

Next we demonstrate how these concepts are used by way of a case study.

7.2.3   Case study: Bluetooth interference aware scheduling (BIAS)

This case study tackles a method for a Bluetooth device to time share the medium with an IEEE 802.11b device. The described previously building blocks are neatly integrated into the specifics of the Bluetooth technology, which is, in this case, a master controlled reservation-based packet transmission system. This algorithm including the discussion of its performance was presented in ref. [42].

In a nutshell, this method allows the master device, which controls all data transmissions in the piconet, to avoid a data transmission to a slave experiencing a "bad" frequency. Furthermore, since a slave transmission always follows a master transmission, using the same principle, the master avoids receiving data on a "bad" frequency by avoiding a transmission on a frequency preceding a "bad" one in the hopping pattern, as illustrated in Figure 7.1.

This technique relies on a channel estimation procedure that can classify frequencies as "good" or "bad" using criteria such as packet loss, negative ACK, or any other technique presented in Chapter 6. In addition, the scheduling scheme illustrated

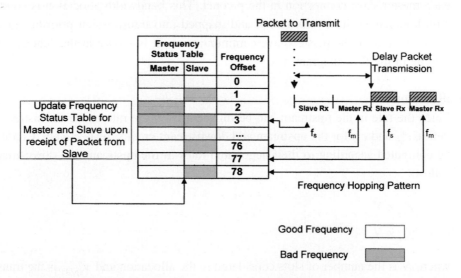

Fig. 7.1. Interference aware scheduling (taken from ref. [42]).

in Figure 7.1 needs only to be implemented in the master device and translates into the following transmission rule. The master transmits in a slot after it verifies that both the slave's receiving frequency, f_s, and its own receiving frequency, f_m, are "good". Otherwise, the master skips the current transmission slot and repeats the procedure during the next transmission opportunity. In the remainder of this section, we first discuss the transmission rules for the master and the slave device. Then we discuss a credit allocation function and a service priority routine that schedules packet transmissions to devices according to their service requirements and the state of the channel.

Transmission rules

Let us establish the following transmission rules for the master and the slave devices.

The master polls the slave, S_i, every p^i slots in order to guarantee γ^i_{up} in the upstream direction. A poll message can be either a data or POLL packet. A data packet is sent if there is a packet in the queue destined for S_i. This packet contains the ACK of the previous packet received from S_i. If there are no data to transmit and the master needs to ACK a previous slave transmission, it sends a POLL packet.

Upon receipt of a packet from the master, the slave can transmit a data packet. This data packet contains the ACK information of the master to slave packet transmission. If the slave does not have any data to send, it sends a NULL packet in order to ACK the previous packet reception from the master. No ACK is required for a NULL message from the master.

Figure 7.2 describes the master's transmission flow diagram. In addition to checking the slave's and the master's receiving frequencies pair, (f_s, f_m), the algorithm incorporates bandwidth requirements and quality of service guarantees for each master/slave connection in the piconet. This bandwidth allocation is combined with the channel state information and mapped into transmission priorities given to each direction in the master/slave communication. It is shown in the "choose slave" routine in the flow diagram.

Credit allocation

Since the rate in the upstream can be different from the rate in the downstream, we define c^i_{up} and c^i_{dn} for the upstream and downstream credits, respectively. Credits can be computed according to their negotiated rates in the upstream and downstream as follows:

$$c^i_{up} = \gamma^i_{up} \times N \tag{7.6}$$
$$c^i_{dn} = \gamma^i_{dn} \times N$$

where N is the number of slots considered in the allocation and $\gamma^i_{up/dn}$ is the transmission rate in the upstream/downstream. Given the packet size in the up/downstream,

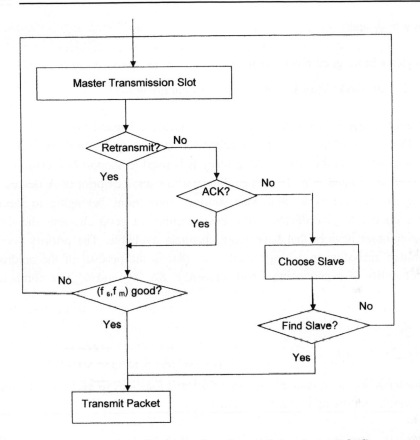

Fig. 7.2. Master packet transmission flow diagram (taken from ref. [42]).

$l_{\text{up/dn}}^i$, and the number of back-to-back packets sent during a poll interval p^i, $\gamma_{\text{up/dn}}^i = l_{\text{up/dn}}^i (N_{\text{peak}}^i / p^i)$. Note that credits are decremented by the number of slots used in each data packet transmission; γ is derived from the application QOS parameters such as delay, peak bandwidth, and jitter. For more details on how to compute γ, the reader is referred to ref. [42].

7.2.4   Service priority

The third component of the algorithm is to give an access priority to certain devices based on the channel conditions and their allocated credits.

Let u_i be the probability that a pair of master/slave transmission slots are "good". Thus, u_i represents the available spectrum to the slave S_i, and we write

$$u_i = \min((1 - \epsilon), P(\text{slave } i \text{ has a good receiving frequency})$$

$$\times P (\text{master has a good receiving frequency})) \tag{7.7}$$

where $\epsilon \approx \frac{1}{79}$ and

P (device i has a good receiving frequency)

$= $ number of good channels$_i$/total number of channels   (7.8)

We use a two-tier system with high and low priorities, denoted by A and B, respectively. Priority A is used to support delay constrained applications such as voice, MP3, and video. On the other hand, priority B is used to support best effort connections such as ftp and http. The scheduling routine services priority A devices first and priority B devices second. Also, among connections belonging to the same tier, we choose to give devices with fewer number of good channels the right of way over other devices that have more channels available. The priority access is determined according to a weight factor, w, that is the product of the credits and the probability of experiencing a bad frequency. So, w_{up}^i and w_{dn}^i are computed as follows:

$$w_{up}^i = c_{up}^i \times (1 - u_i)$$
$$w_{dn}^i = c_{dn}^i \times (1 - u_i)$$
(7.9)

The master schedules a data transmission for slave i such as to maximize the product of the weights in the up and downstreams:

$$i = \max_S^f (w_{up}^i \times w_{dn}^i)$$
(7.10)

To transmit a POLL packet in the absence of data in the downstream, and in order to allow for a slave to transmit its data, the master looks only at the weight function in the upstream:

$$i = \max_S^f (w_{up}^i)$$
(7.11)

The selection of a slave is restricted over the set of slaves S that can receive on the master's current transmission frequency, f. Thus, any slave that experiences a high level of BER on the current transmission frequency is not considered. Four sets of slaves are formed, A_{data}^f, A_{poll}^f, B_{data}^f, and B_{poll}^f. Note that A_{data} and A_{poll} represent the set of high priority connections requiring data and POLL packet transmissions, respectively. Similarly, B_{data} and B_{poll} represent low priority connections. First, the algorithm tries to schedule a packet to high priority slaves in group A, then a POLL packet, before it moves to group B. The credit counters and weights are updated accordingly after every master's transmission. Table 7.1 summarizes the parameters used in the algorithm and their definition, and Table 7.2 gives the algorithm's pseudo-code.

Table 7.1. *Definition of parameters used in the scheduling algorithm*

Parameters	Definition
$\gamma^i_{\mathrm{up,dn}}$	rate allocated for device i in the upstream and downstream
$w^i_{\mathrm{up,dn}}$	weight for device i
$c^i_{\mathrm{up,dn}}$	credit for device i
N	number of slots considered in the allocation
u^i	available frequency usage for device i

Taken from ref. [42]

Table 7.2. *BIAS pseudo-code*

1:Estimate_channel(); // Perform (offline or online) estimation
2:Every N Slots
3:    compute_credits();
4:Every Even TS_f // master transmission slot
5:    if $TS_f + l_{\mathrm{dn}}$ is clear   // master can receive on next slot
6:    {
7:        A^f_{data} = {set of high priority slaves such that
                    ((f "good") and ($qsize > 0$) and ($c_{\mathrm{dn}} > 0$) }
8:        A^f_{poll} = {set of high priority slaves s.t. ((f "good") and ($c_{\mathrm{up}} > 0$)) }
9:        B^f_{data} = {set of low priority slaves s.t. ((f "good") and ($qsize > 0$)) }
10:       B^f_{poll} = {set of low priority slaves s.t. ((f "good") and ($c_{\mathrm{up}} + c_{\mathrm{dn}} > 0$)) }
11:       // Service high priority slaves first
12:       if ($A^f_{\mathrm{data}} \neq$ empty)   // transmit data packets
13:       {
14:           $i = \max_{A^f_{\mathrm{data}}} (w^i_{\mathrm{up}} \times w^i_{\mathrm{dn}})$ // select device i with the largest weight
15:           transmit data packet of size l_{dn} to slave i
16:       }
17:       else if ($A^f_{\mathrm{poll}} \neq$ empty)   // transmit polls
18:       {
19:           $i = \max_{A^f_{\mathrm{poll}}} (w^i_{\mathrm{up}})$ // select device i with the largest weight
20:           transmit poll to slave i
21:       }
22:       // Then service low priority slaves
23:       else if ($B^f_{\mathrm{data}} \neq$ empty)
24:       {
25:           $i = \max_{B^f_{\mathrm{data}}} (w^i_{\mathrm{up}} \times w^i_{\mathrm{dn}})$ // select device i with the largest weight
26:           transmit data packet of size l_{dn} to slave i
27:       }
28:       else if ($B^f_{\mathrm{poll}} \neq$ empty) // transmit polls
29:       {
30:           $i = \max_{B^f_{\mathrm{poll}}} (w^i_{\mathrm{up}})$ // select device i with the largest weight

Table 7.2. (*cont.*)

31:	transmit poll to slave i
32:	}
33:	}
34:	update_credits();
36:	// high priority slaves
37:	$c_{up}^i = max(0, c_{up}^i - l_{up}^i)$ /hspace*0.1in//upstream packets
38:	$c_{dn}^i = max(0, c_{dn}^i - l_{dn}^i)$ /hspace*0.1in//downstream packets
39:	// low priority slaves
40:	if $(c_{up}^i > 0)$ $c_{up}^i = max(0, c_{up}^i - l_{up}^i)$ //upstream packets
41:	else $c_{dn}^i = max(0, c_{dn}^i - l_{up}^i)$
42:	if $(c_{dn}^i > 0)$ $c_{dn}^i = max(0, c_{dn}^i - l_{dn}^i)$ //downstream packets
43:	else $c_{up}^i = max(0, c_{up}^i - l_{dn}^i)$
44:	$w_{dn,up}^i = (1 - u_i) \times c_{dn,up}^i$ // update weights

Taken from ref. [49]

Now that we have discussed each component of BIAS, let us summarize some of its main properties:

(P1) Service guarantees are provided to error free connections including delay bounds and throughput. Although error prone connections are given a higher priority of access, interference free devices are not affected by the interference conditions subsiding on some devices in the piconet.

(P2) The scheduling policy is work conserving since no slots will be left idle if there is at least one device with a positive credit counter.

(P3) The sharing of "good" channels is proportional to each session's negotiated rate regardless of whether they are error-free or error-prone.

Further details and proofs of the fairness properties listed above are found in ref. [42].

Performance evaluation results

Now let us evaluate the performance of the algorithm given in Table 7.2. The experiment we use highlights the support of QOS in an environment in which devices experience different levels of interference and connections have a wide range of service requirements.

We use the network topology illustrated in Figure 7.3. Slaves 1 and 2 experience the same level of interference, while slave 3 does not experience any interference. The y-coordinate of the WLAN mobile device is varied along the y-axis in order to vary the level of interference on the Bluetooth piconet.

The scenario includes three different application profiles, namely video, print, and email. The video application has strict delay constraints and tolerance to packet loss. On the other hand, the print and email applications are more delay and loss tolerant.

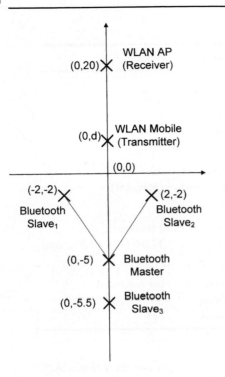

Fig. 7.3. Multislave Bluetooth piconet and two WLAN nodes.

In order to stress the evaluation environment, the device running the video application is placed closer to the WLAN interference source that is running an FTP connection. The application profiles are summarized in Table 7.3.

The results obtained when using BIAS are compared with those using round robin scheduling (denoted by RR). The results are summarized as follows. BIAS keeps the delay of the video application the same (compared to RR), while it reduces the packet loss to less than 0.5% (compared to 15% with RR). In addition, the delays for the print and email applications are lowered by a few milliseconds, and the packet loss is decreased by at least an order of magnitude.

Before discussing the results, we first show how we set the transmission rates in the upstream, γ_{up}, and downstream, γ_{dn}.

With the exception of voice and video traffic, applications such as FTP, HTTP, print and email do not have delay constraints. Therefore, in our experiment the video application represents the high priority traffic, while the print and the email applications are low priority.

We divide the bandwidth equally among the high and low priority traffic. We also verify that 50% of the bandwidth is sufficient for the video application to transmit data and that packets do not accumulate in the buffers. Therefore, we set $\gamma_{up}^2 = \gamma_{dn}^2 = 0.25$. The print and email applications share the lower priority bandwidth,

Table 7.3. *Application profile parameters*

Parameters	Distribution	Value
Email		
Send interarrival time (s)	exponential	120
Receive interarrival time (s)	exponential	60
Email size (bytes)	exponential	1024
Print		
Print requests interarrival time (s)	exponential	30
File size	normal	(30K, 9M)
Video		
Frame rate	constant	1 frame/s
Frame size (bytes)	constant	17280 (128×120 pixels)
WLAN FTP		
Traffic directionality	client to server	put command
Inter-request time (s)	exponential	5
file size (bytes)	exponential	5M

Table 7.4. *Experiment summary*

Node	Application type	γ_{up}^i	γ_{dn}^i
Slave 1	print	0.125	0.125
Slave 2	video	0.25	0.25
Slave 3	email	0.125	0.125

i.e. $\gamma_{up/dn}^1 = \gamma_{up/dn}^3 = 0.5/4 = 0.125$. Table 7.4 details the parameters used in this experiment.

Figures 7.4(a) and (b) give the access delay for each application measured at the slave and the master, respectively.

Observe in Figure 7.4(a), that the delay with BIAS is lower than that with RR for all three slaves. The difference ranges from 200 ms for slave 1 to a few milliseconds for slaves 2 and 3.

The delay in the upstream shown in Figure 7.4(b), is consistent with the results in the downstream. We observe lower delays for slaves 1, 2, and 3 by about 0.01, 0.2, and 0.2 s, respectively. This depends on the exact traffic distribution, the number of packets in the buffer awaiting transmission, and the packet loss level. In the downstream, by delaying the transmission of a packet, a retransmission following a packet collision is avoided, and thus the access delay is reduced.

Figure 7.5 gives the packet loss at both the master and the slave with BIAS and RR. Given the placement of slave 3 far from the WLAN interference source, the packet loss for slave 3 is extremely low, in the order of 10^{-3}–10^{-5} with RR. We also observe two distinct group of curves. With BIAS the packet loss for slaves 1, 2, and the master is between 0.001 and 0.01. It is at least two orders of magnitude higher than with RR.

In summary, the results of this experiment and others conducted in order to calibrate the parameters used of the bandwidth allocation lead us to the following two observations.

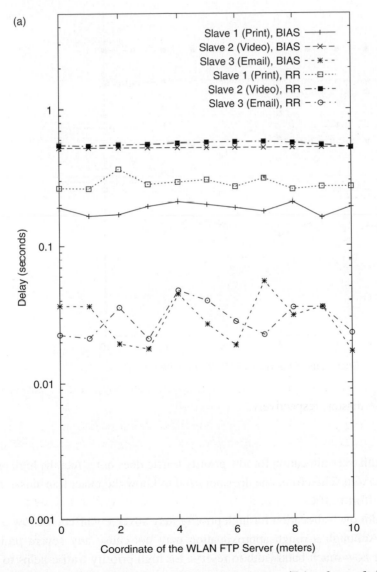

Fig. 7.4. (a) Delay at the slave. (b) Delay at the master. (Taken from ref. [42]).

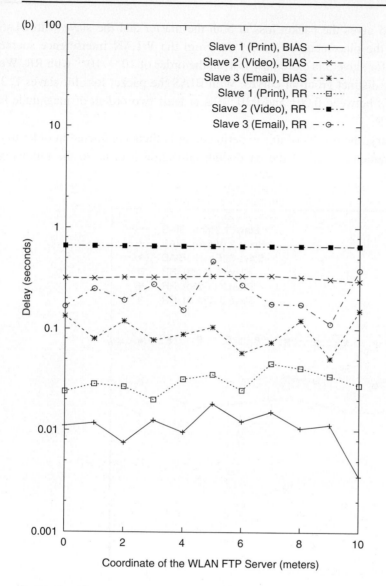

Fig. 7.4. (cont.).

(i) Bandwidth over-allocation for low priority traffic does not affect the high priority traffic service. Therefore, one does not need to know the exact transmission rates of best effort traffic.

(ii) Bandwidth over-allocation for high priority may adversely affect the low priority traffic. Although a rough approximation may not cause any severe problems, knowing how much bandwidth to reserve for high priority traffic helps to avoid bandwidth waste.

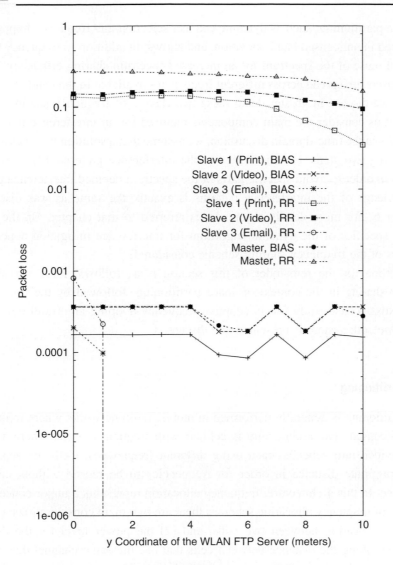

Fig. 7.5. Probability of packet loss. (Taken from ref. [42]).

7.3   Frequency division multiple access

Frequency sharing techniques constitute a major part of the interference mitigation solution space. Frequency sharing mechanisms are usually combined with either space or time partitioning. Space partitioning is mostly popular in the context of cellular networks where a narrowband channel is assigned to a specific geographic area and all users in that area are assigned to the same frequency. Narrowband channels are spatially reused in order to overcome spectrum scarcity. On the other hand, time partitioning, such as dynamic channel selection and frequency hopping, is mainly used in unlicensed band operation, and allows, in addition to frequency reuse, a temporal reuse of the spectrum for an increased spectrum sharing efficiency. Thus, different users are spread across different frequencies available in the band over time.

Before we discuss in greater detail these two flavors of frequency sharing mechanisms, let us consider the main components required for an interference mitigation solution. As in the time-domain discussion, we assume that a solution involves at least the following two steps: (1) knowledge of the interference patterns, (2) a dynamic algorithm in order to avoid specific parts of the spectrum deemed interference prone. The knowledge of the interference patterns is exactly the same as was discussed in Chapter 6. For more details, the reader is referred to that chapter. On the other hand, the specifics of the dynamic algorithm for interference mitigation depend on the flavors of the frequency sharing scheme considered.

The outline for the remainder of this section is as follows. First, we discuss frequency sharing in the context of space partitioning, followed by the time partitioning. Next, a case study for an adaptive frequency hopping algorithm developed for the Bluetooth system is presented and discussed.

7.3.1   Space partitioning

Space partitioning is generally performed in mobile radio networks where frequency usage is licensed. The architecture is cellular with frequency reuse, where a large area is divided into subcells, each using different frequencies. Cells are separated by an appropriate distance in order for frequencies to be reused without causing interference. In this architecture, frequency allocation represents a major concern. In the design of frequency allocation schemes there are two main considerations: (1) the geographical location between two cells, and (2) the power level for the desired signal. Controlling the distance between cells that use the same channel determines the level of co-channel interference obtained. Two cells can use the same channel only when the geographical distance between them is greater than D, as depicted in Figure 7.6. Otherwise, their communication session suffers from the so-called channel interference. Thus, there is a trade-off between achieving a high level of channel reutilization and not exceeding a signal to interference ratio threshold. The other concern is adjusting the power level of the transmitter in order to increase the desired signal power level while reducing the interference caused on adjacent nodes. Channel allocation techniques for cellular networks are divided into three categories: fixed, dynamic, and hybrid allocation. In fixed allocation, channels are allocated in the network deployment phases based on the expected number of users, traffic load, cell coverage area, etc. In dynamic channel allocation, the channels are assigned by the mobile switching center on a regular basis considering channel reuse distance, future

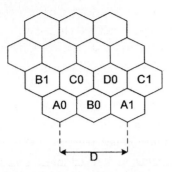

Fig. 7.6. Space partitioning.

call blocking probability, and channel reuse frequency. Hybrid allocation combines both methods.

7.3.2   Time partitioning

It is common to use time partitioning in conjunction with frequency division multiple access. The time partitioning granularity used varies from a symbol to a packet interval. In Chapter 2, we discussed fast frequency hopping systems versus slow frequency hopping systems. Fast frequency hopping systems refer to several symbols transmitted over several frequencies, while slow frequency hopping relates to several symbols or a packet transmitted over a single frequency. In some other cases, several packets are transmitted over the same frequency, before a new channel is selected. This is commonly referred to as dynamic channel selection.

In the context of this discussion, we use the term frequency hopping to designate all time granularities available, although we are mostly concerned with the packet level time granularity corresponding to slow frequency hopping and dynamic channel selection.

As we stated previously, the way frequency hopping radios work makes them more resilient to interference. Let us look at a practical example in order to find out how this interference resilience is achieved.

Basically, hopping is performed across multiple fixed frequencies where each packet is sent on a different frequency. Typically, the hopping pattern is pseudo-random and repeats depending on the size of the frequency band considered. When an interference signal causes a packet loss on a particular frequency, the frequency hopping system has already moved to the next frequency in the hopping pattern and therefore should be able to avoid it the next time around. In other words, an interfering signal may lead to data loss, but due to the randomness of the hopping patterns, some data should still get through regardless of the interference power level. This enables a frequency hopping radio to transmit reliably small packets potentially through areas of heavy interference. This is illustrated in Figure 7.7, which shows

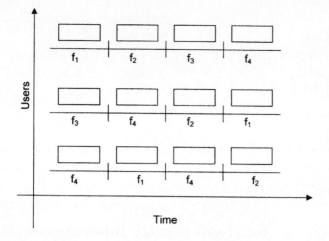

Fig. 7.7. Time partitioning.

how several users make use of time and frequency partitioning in order to avoid interfering with each other.

While most frequency hopping systems available today use either a random hopping pattern or a pre-selected pattern, the key is how to modify adaptively the hopping patterns in order to mitigate interference.

A simple rule that can be implemented in an adaptive algorithm consists of the following:

1: If hop_f = "bad"   // the frequency corresponding to this hop is bad
2:     Frequency_pattern − hop_f // remove f from hopping pattern;

In the next section, we discuss an example algorithm developed for the Bluetooth system.

7.3.3   Case study: adaptive frequency hopping for Bluetooth systems

The key idea in adpative frequency hopping (AFH) is to eliminate the so-called "bad" frequencies or, in other words, replace "bad" frequencies with "good" ones so that transmissions are free from interference. A discussion for this algorithm and its performance appeared in refs [46] and [47].

In the case study we discussed for time division multiplexing, the main idea was to time share the medium, so that transmissions occuring in "bad" frequencies are postponed. Obviously, in this case, while interference may be avoided, there are still inefficiencies since slots associated with "bad" frequencies have to be skipped. In essence, a part of the bandwidth is not available to some devices. On the other hand, in AFH we look to exploit frequency division multiplexing in order to avoid hopping on "bad" frequencies. By randomizing the hopping pattern, random noise and other frequency hopping systems can be avoided. In addition, by adapting the hopping pattern, users of fixed bandwidth can also be avoided.

First, we describe the Bluetooth frequency hopping sequence defined in the Bluetooth specifications [1], then we present an AFH algorithm that modifies it in order to mitigate interference.

Frequency hopping in Bluetooth is achieved as follows. Frequencies are sorted into a list of even and odd frequencies in the 2.402–2.480 GHz range. A segment consisting of the first thirty-two frequencies in the list is chosen. After all thirty-two frequencies in that window are visited once in random order, a new window is set including sixteen frequencies of the previous window and sixteen new frequencies in the sorted list. From the many AFH algorithms possible, here is an implementation that eliminates "bad" frequencies in the sequence.

```
1: W_FH = segment_size;          // Initialize the hopping algorithm window size
2: W_FH+ = N_BF;                 // Increase by the number of "bad" frequencies
3: If (W_FH > 79)
4:     W_FH = 79;                // limit to the list size
5:         if (N_BF ≥ 79 − min_hop)
6:             //use min_hop "good", "bad" frequencies
```

1: $W_{FH} = segment\_size$;   // Initialize the hopping algorithm window size
2: $W_{FH}+ = N_{BF}$;   // Increase by the number of "bad" frequencies
3: If ($W_{FH} > 79$)
4:     $W_{FH} = 79$;   // limit to the list size
5:       if ($N_{BF} \geq 79 -$ min_hop)
6:       //use min_hop "good", "bad" frequencies

Given a segment of thirty-two "good" and "bad" frequencies, the algorithm visits each "good" frequency exactly once. Each "bad" frequency in the segment is replaced with a "good" frequency selected from outside the original segment of thirty-two, as shown in Figure 7.8.

Thus, the difference between AFH and the original Bluetooth hopping sequence algorithm is in the selection of only "good" frequencies in order to fill up the segment size. Some additional constraints can be imposed on the maximum number of "bad" frequencies to be eliminated if a minimum number of different frequencies is to be kept in the sequence. In their most recent ruling, the FCC recommends using at least fifteen different frequencies.

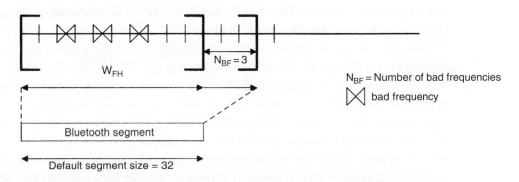

Fig. 7.8. Bluetooth adpative frequency hopping patterns (taken from ref. [46]).

Changing the frequency patterns requires changes in the Bluetooth hardware implementations. Another requirement is the advertisement of the new hopping pattern among devices in the piconet in order to retain synchronization. This is typically achieved using link management protocol (LMP) messages exchanged between the master and the slaves in the piconet in order to advertise the new hopping sequence. This last requirement imposes some limitations on how often a new hopping pattern should be advertised and used. Improving performance such as lowering the packet loss, the access delay, and increasing the throughput should outweigh the communication overhead associated with synchronization. The synchronization update interval could be dynamically adjusted so that it tracks the channel dynamics.

Finally, AFH does not preclude additional scheduling techniques to control the transmission (and possibly the retransmission) of packets on the medium.

Performance evaluation results

Now let us study the performance of the AFH mechanism discussed previously. We consider a set of experiments where we vary the application considered, as this factor is most likely to dominate the performance results. The discussion for these results appeared in ref. [47].

The network topology studied consists of four nodes: one WLAN mobile node connected to a WLAN access point, and a Bluetooth piconet including one master and one slave device. The Bluetooth piconet is 1 m away from the WLAN mobile node. The WLAN access point is 15 m away from the Bluetooth piconet.

For Bluetooth, we consider two applications: FTP and voice. FTP is a bandwidth hungry application that stresses the throughput requirement, while voice has strict delay and jitter requirements. Together, these two applications constitute a representative set of the application space used in a Bluetooth piconet. For WLAN, we use FTP to upload a large file (for instance a movie) to a server.

For the FTP profile, the parameters are the traffic directionality, the inter-request time, and the file size. For Bluetooth we vary the file sizes from 200 bytes to 500 kbytes (every 5 s), while for WLAN we use a single file of 960 Mbytes. The voice application used in Bluetooth is based on the G.723.1 encoder. The profile parameters are summarized in Table 7.5.

Now we discuss the details of two experiments involving a voice and an FTP application for Bluetooth and an FTP application for WLAN. In experiment 1, we vary the file size of the Bluetooth FTP application. A summary of the experiments is given in Table 7.6.

In experiment 1, we consider the effects of the AFH scheme on the performance of a Bluetooth FTP connection when it is operating in close proximity to a WLAN FTP connection. While the WLAN connection is used to upload a 960 Mbyte file to a server, a Bluetooth FTP connection is used to transfer files between two devices equipped with a Bluetooth radio, for example a PDA or a laptop. This latter operation

Table 7.5. *Application profile parameters*

Parameters	Distribution	Value
Bluetooth FTP		
Traffic directionality	server to client	get command
Inter-request time (s)	exponential	5
File size (kbytes)	variable	[0.2, 500]
WLAN FTP		
Traffic directionality	client to server	put command
File size (Mbytes)	constant	960
Bluetooth voice		
Encoder		G.723.1
Silence length (s)	exponential	0.65
Talk spurt (s)	exponential	0.352

Table 7.6. *Experiment summary*

Experiment	Bluetooth application type	WLAN application type
1	FTP	FTP
2	voice (G.711)	FTP

produces similar traffic characteristics to that of a "HOT SYNC", even if the file sharing protocol used in that case is specific to the PDA manufacturer.

Figure 7.9 gives the packet loss results at the Bluetooth slave device; RR refers to the case when a round robin algorithm is used, while AFH refers to the use of AFH. When no scheme is used, the packet loss starts at 12%. On the other hand, the packet loss for AFH starts at 2% and increases to 6% as the offered load is increased to 800 kbit/s. Note that the packet loss observed with AFH depends on the frequency of the synchronization messages exchanged between the Bluetooth master and the slave. There is a trade-off between the communication overhead and the response to changes in the interference environment. A fast responding system will incur a lower packet loss at the cost of a higher communication overhead. In this experiment, synchronization messages are exchanged on average every few seconds (two to ten).

Figures 7.10(a) and (b) illustrate the TCP goodput and delay results, respectively. We observe that the goodput is directly proportional to the offered load until about 480 kbit/s for both curves. We have computed that about 600 kbit/s is the maximum application goodput available considering the choice of the experiment parameters. This includes a 10% overhead for the packet headers of all layers between the application and the Bluetooth baseband link and assuming a maximum TCP packet payload of 1460 bytes. Thus, 480 kbit/s corresponds to 80% of the Bluetooth medium

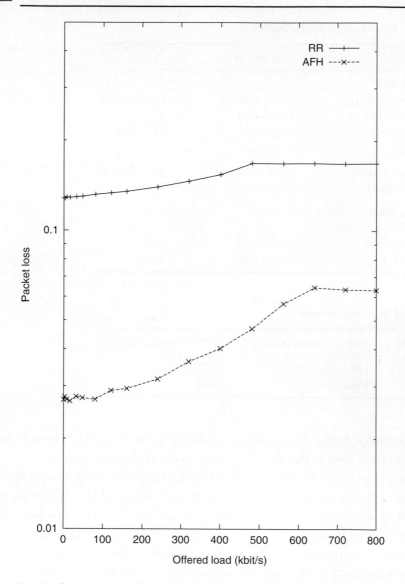

Fig. 7.9. Experiment 1: Bluetooth packet loss.

capacity. As the offered load is increased beyond 500 kbit/s, the difference between the various schemes becomes more significant. The maximum goodput obtained is 600 kbit/s with AFH, while it is only 480 kbit/s when no algorithm is used.

The TCP file transfer delay shown in Figure 7.10(b) is consistent with the goodput results. The file transfer delay remains below 4 s until 500 kbit/s for AFH. It is 2 s higher when no algorithm is used. Both delay curves take off sharply when the offered load is increased above 500 kbit/s.

In summary, AFH improves the maximum Bluetooth goodput by 25%. It is important to point out that in this experiment the interference level remains the same for several minutes since the WLAN connection is transmitting during the entire simulation time. Therefore, the communication overhead associated with having to adapt the frequency sequence is low, and AFH gives the highest Bluetooth throughput. Had the WLAN traffic been more bursty, additional packet loss could have been incurred with AFH, and the throughput advantage may not have been as significant.

While in experiment 1 the objective is to maximize the throughput of the FTP connection, in experiment 2 the goal is the minimize the delay, and most importantly the delay jitter, for a Bluetooth voice connection. We use the same parameters as used

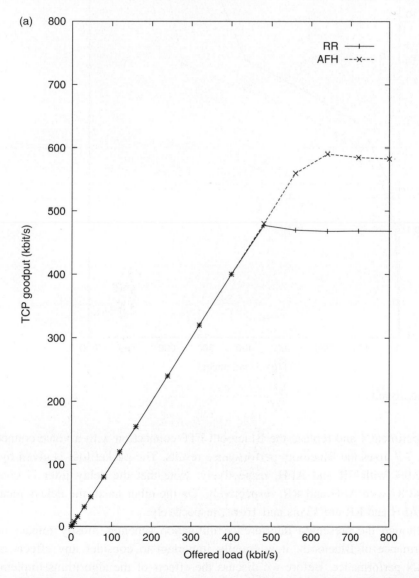

Fig. 7.10. Experiment 1. (a) Bluetooth goodput (kbit/s); (b) Bluetooth TCP delay (seconds).

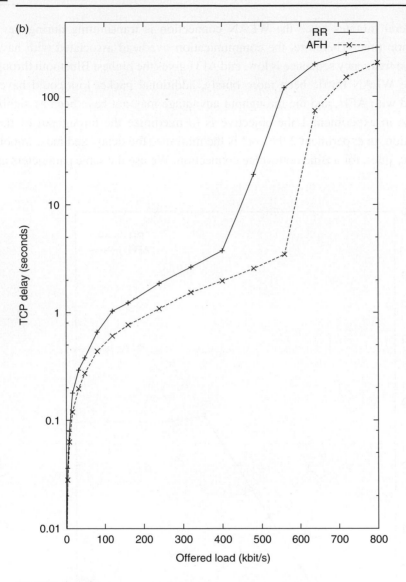

Fig. 7.10. (cont.).

in experiment 1 and replace the Bluetooth FTP connection with a voice connection. Table 7.7 gives the Bluetooth performance results. The packet loss is given by 11% and 2.9% with RR and AFH, respectively. Note that the delay jitter is given by 82 and 87 with AFH and RR, respectively. On the other hand, the delays measured with AFH and RR are 13 ms and 16 ms, respectively.

Although the presented interference mitigation schemes mostly impact on the performance of Bluetooth, it is equally important to consider any effects on the WLAN performance. Before we discuss the effects of the algorithms implemented for Bluetooth on the WLAN, it is important to keep in mind that in the simulation

Table 7.7. *Experiment 2: Bluetooth voice performance*

	AFH	RR
Packet loss	0.0298	0.1103
Delay (s)	0.0013	0.0016
Delay jitter (s)	0.0827	0.0879
Goodput (kbit/s)	541.536	509.644

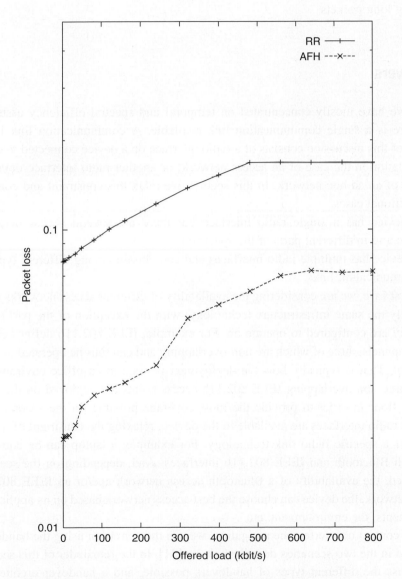

Fig. 7.11. Experiment 1: WLAN probability of packet loss.

setup used, the WLAN node that is close to the Bluetooth piconet is mainly functioning as a transmitter of data packets and not a receiver. Thus, the impact of the Bluetooth interference is not as significant since the WLAN node only receives short ACK packets. Figure 7.11 shows the WLAN packet loss observed on the WLAN receiver located on the laptop computer. When no interference mitigation algorithm is implemented for Bluetooth, the packet loss is 17%. The packet loss when AFH is implemented drops to 7%. Note that we expect the packet loss to be more significant with None and AFH (up to 30% and 15%, respectively) when the WLAN node is receiving long packets.

7.4   Handovers

So far we have mostly concentrated on temporal and spectral efficiency assuming that there is a single communication link available. A communication link in the context of this discussion consists of a radio interface on a device connected to either a base station in the case of an access network, or another radio interface device in the case of an ad hoc network. In this section, we relax this constraint and consider two additional cases.

(i) A device has a single radio interface but there are several access networks operating in different parts of the spectrum.

(ii) A device has multiple radio interfaces and can choose among different types of communication links.

In the first case, we are considering the availability of different access networks using essentially the same infrastructure technology, with the exception of the part of the band they are configured to operate on. For example, IEEE 802.11b defines eleven center channels, three of which are non-overlapping and can thus be operated in close proximity. That is typically how the deployment is done in an office environment, where three non-overlapping IEEE 802.11b access points are deployed on the same building floor in order to provide the most coverage possible. In the second case, different radio interfaces are available to the device, relaxing the constraint of having to choose a specific radio link technology. For example, a laptop can be equipped with both Bluetooth and IEEE 802.11b interfaces. And, depending on the scenario considered, the availability of a Bluetooth access network and/or an IEEE 802.11b access network, the device can choose the best access network based on its application requirements, the environmnent, etc.

In the context of interference mitigation, we can then envision using the handovers described in the two scenarios described above [31]. In the remainder of this section, we discuss the different types of handovers possible, and a handover architecture currently being developed by different standards organizations in order to optimize and facilitate handovers.

7.4.1   Handover types

There are different types of handovers, namely horizontal, vertical, multiple interface management, and multiple flow management.

(i) Horizontal handovers are constrained to a single interface and are usually initiated when a mobile device leaves the administrative boundaries and the coverage area of the access network it is currently connected to [53].

(ii) Vertical handovers are performed between multiple interfaces but use a single interface at a time. Once the session is switched to a different interface, the connection resumes on the second interface. In some cases traffic may be sent redundantly on both interfaces until the handover is complete. The mobile device does not need to move in order for this handover to be performed.

(iii) Multiple interface management represents the simultaneous use of different interfaces, one for each traffic flow or session. There is typically an association between the application and the interface used.

(iv) Multiple flow management represents the ability to split flow among different network interfaces according to quality of service requirements and user preferences.

The first two types of handover are particularly interesting in the context of interference mitigation, since they allow one either to change the association with an access point that may be operating in a "bad" frequency band, or switch to another interface. These two scenarios are depicted in Figure 7.12.

Now that we know the types of handovers that could be used in interference mitigation, we turn to handover implementation considerations.

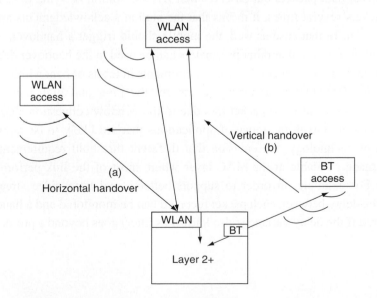

Fig. 7.12. Horizonal versus vertical handovers.

7.4.2   Handover architecture

The proliferation of access network technologies and their deployment has created an unprecedented need to combine different types of networks that users can roam "seamlessly" without experiencing any connection disruption. This need for a seamless mobility is putting the various types of handovers at the center of protocol development and standardization. Since most existing protocols do not meet the user's expectation for seamless mobility, there is a need to extend and optimize protocols in order to improve handover performance. The development of a generalized handover architecture has started recently, mainly led by efforts in the IEEE 802 and the IETF Working Groups.

This architecture consists of several components, namely handover triggers, policy, handover support functionality, including signaling and information management protocols carried over layer 3 protocols [12,13].

In an interference constrained environment, several layer 1 and 2 performance measures can be used to trigger a handover between two access networks. Since the signal level measure alone may not necessarily reflect the level of interference between the victim device and other technologies operating in the same band, we can envision the use of a number of packets retransmitted at the MAC layer as a measure of the number of packets that have been discarded due to packet collisions. As the number of retransmissions exceeds a threshold value, the mobile device may trigger a handover. Note that this measure captures the percentage of packet loss at the receiver. In addition, this measure can indirectly provide information about packet collisions at the receiver at the other end. For example, if a mobile device correctly receives data packets but observes that its base station is trying to send the same data packets several times, it means that the station's ackowledgments are lost at the base station. In that case as well, the mobile should trigger a handover.

At this point we observe that other parameters can be used in the handover decision in order to capture the specific application requirements in terms of bandwidth, delay, and packet loss. If real time applications require low jitter and high bandwidth, TCP traffic is more sensitive to packet loss due to the window congestion algorithm and flow adaptation. Given the range of applications that are likely to be supported in different link technology, we envision that different threshold requirements are devised and made available at the MAC layer where most of the link performance is measured. For example, in order to support better a real time video streaming application, the delay between each packet received can be monitored and a handover can be triggered if the delay variance (also known as jitter) goes beyond a pre-defined threshold.

8 Myths and common pitfalls

This chapter's main objective is to discuss and to some extent dispel some common myths and misconceptions associated with interference mitigation solutions. Our goal is to shed some light on the lessons learned while researching and developing solutions.

A common path taken in the development of interference mitigation techniques often begins by identifying solutions developed for different purposes and applying to the problem at hand. In general, this path constitutes an extremely powerful approach, and examples given in Chapter 7, including time and spectral multiplexing, clearly demonstrate the effectiveness of the resulting solutions. However, applying solutions out of the original context for which they were developed is no simple task, since it requires a careful examination of all the parameters and the assumptions that come into play. It is often when this step is overlooked that myths are constructed.

Contrary to common belief, we show that some techniques often associated with interference mitigation do not constitute solutions. These techniques may in fact aggrevate the interference problem or have a negative impact on the overall system performance. They constitute what we call pitfalls that should be avoided if possible.

We find two recurring myths in most pitfalls studied, although this list is far from exhaustive.

(i) Dealing with interference is similar to dealing with random noise and other wireless channel propagation properties and impairments.

(ii) A set of system parameters such as transmitted power, offered load, packet size, error correction scheme, and modulation techniques can be optimized in order to mitigate interference.

We observe that these two themes are intimately intertwined, and it is no coincidence that most pitfalls that we will dicuss generally belong to both themes.

We now proceed to the discussion of select pitfalls. Our discussion will include several illustrative examples consisting of specific scenarios, parameters, technologies, and performance results wherever applicable. While the absolute numbers presented have little value outside the context in which they are used, they make it possible to anticipate general performance trends and trade-offs.

Finally, by pointing out common pitfalls we hope to guard against hasty applications of solutions without careful consideration of all the assumptions and conditions involved.

8.1   Power control

Power control (PC) techniques have always been considered at the heart of interference mitigation. We find the application of power control techniques in most systems today to combat channel propagation issues, noisy environments, and even interference in the case of CDMA systems.

Given that most devices provide the ability to modify dynamically their transmission power, it is a good idea to investigate the dynamics of a PC strategy as a means of alleviating the impact of interference. A couple of observations are in order. Reducing transmission power decreases the power consumption of the transmit mode and has also the advantage of reducing the interference noise level for neighboring devices. The disadvantage, however, is that when a device reduces its transmission power, it decreases the signal to interference ratio of its transmission, thus increasing its bit error rate. Another important issue to investigate is the dynamics of a responsive system. For example, let us consider two pairs of communicating devices, $A_{1,2}$ and $B_{1,2}$. If the device pair $A_{1,2}$ decides to raise its transmitted power due to a low measured SIR, that may prompt device pair $B_{1,2}$ also to increase its transmitted power. This could easily lead to a race situation, where all devices end up using more transmitted power while none achieve any performance gain.

Formally, we can define the transmitted power requirements for each transmitting device in terms of a desired threshold value of SIR, τ, and the given path loss and topology conditions. This leads to a system of equations where each device's transmitted power is unkown [28]. In fact, if we assume the path loss can be computed from any node to any other node, we can define the path loss matrix L_{ij}, where entry ij represents the path loss from node i to node j. Furthermore, we consider nodes i, m, and j_n, as illustrated in Figure 8.1.

Let the transmission between nodes i and m be considered the *main signal* while all other transmissions from nodes j_n, $\forall\, n$, be considered as *jamming signals* (from node i's perspective). Also, let SIR_i be the SIR in dB measured at node i. We can write

$$\text{SIR}_i = P_m - L_{mi} - \left(\sum_{j \neq i,m} P_j - L_{ji} \right) \tag{8.1}$$

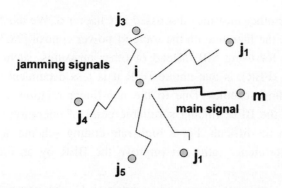

Fig. 8.1. Signal received at node i.

where P_m is the power transmitted by node m and L_{ji} is the path loss in dB on link ji (node j transmitting and i receiving). Given different SIR requirements for different error rates, we let τ_i be the minimum required SIR threshold in order to achieve a certain link quality. Therefore, we can rewrite Equation (8.1) as follows:

$$\text{SIR}_i \geq \tau_i \tag{8.2}$$

Rewriting the inequalities of Equation (8.2) in matrix form yields

$$(I - 1)P + rIL_i \geq \Gamma \tag{8.3}$$

where $P = (P_1, P_2, \ldots, P_i, \ldots, P_N)$ is the column vector of the transmitted powers, I is the identity matrix, $\mathbf{1}$ is the indicator function ($\mathbf{1}_{i \neq j}$ is 1 if $i \neq j$ and 0 otherwise) and Γ is the column vector of the threshold SIR, τ_i, $r = (1, 1, \ldots, -1, \ldots, 1)$ is a $1 \times n$ row vector with all 1's and a -1 in column m, and L_i is column i of the path loss matrix L. Note that L is an $n \times n$ matrix consisting of the path loss values (in dB) from any node i to node j in the system. It is easy to see that $L_{ii} = 0 \; \forall \; i$. The solution P^* depends on whether $(I - 1)$ is invertible. If a solution vector, $P^* = (P_1, P_2, \ldots, P_i, \ldots, P_N)$, exists, it can be computed by solving the system of inequalities given in Equation (8.2).

Now, let us look at a distributed algorithm in order to implement a power control procedure. This distributed algorithm should converge to the solution, P^*, when there is one. The goal is to study whether adapting the transmitted power in an interference environment can lead to any performance improvements. The basic idea is to adjust the interference level in the system to no more than what is needed. We assume that the receiver does not have any knowledge of other systems except for the system it is communicating with. Interference from other systems can be measured in terms of SIR, RSNI, or some of the other measures discussed in Chapter 6. We use SIR in order to keep consistent with the literature on the topic of power control [28,37,66]. We believe that the reasons for using SIR instead of other measures, such as the BER or the frame error rate (FER), is that unlike BER it is less dependent on the utilized modulation and coding schemes. Due to the non-linear relation between the signal to noise ratio and the BER, finding a suitable point of operation for the power control algorithm can be difficult. For a high rate coding scheme, a small increase in the signal to interference ratio can improve the BER by an order of magnitude.

Initially, we set $P_0 = P_{max}$, then at every update interval U, the power at the transmitter, $P(t+1)$, is updated as follows:

$$P(t+1) = \min\left(P_{max}, \max\left(P_{min}, \frac{\tau_t}{\text{SIR}(t)} \times P(t)\right)\right) \qquad (8.4)$$

where $\tau(t)$ is the target SIR and $\text{SIR}(t)$ is based on an average value over many measurements. The power update rule takes into consideration the $\text{SIR}(t)$ statistic measured at the receiver side. The receiver can then relay this information to the transmitter at every update interval U.

8.1.1   Implementing PC in Bluetooth

Now let us look at the implementation of our distributed PC in a Bluetooth system. This example appeared in ref. [43].

Although the exact details of a power control algorithm have been left undefined for the most part, the Bluetooth specifications have included the necessary hooks in the protocol in order to implement a power control algorithm. Furthermore, the Bluetooth specification classifies devices into three power classes, as summarized in Table 8.1.

Class 1 requires power control limiting the transmitted power over 0 dBm, while power control is optional for classes 2 and 3. The specifications suggest that the transmitted power should be adjusted based on RSNI measurements at the receiver. Note that in an interference limited environment, RSNI corresponds to the SIR (assuming that noise is negligible). Furthermore, the specifications define messages as part of the management protocol, the so-called link management protocol (LMP), in order to adjust the power control. The details of LMP message formats related to PC are left as an exercise for the interested reader.

Another implementation issue to consider is the value of the update interval, U. Andersin *et al.* [27] demonstrate that for a system such as GSM, the SIR

Table 8.1. *Bluetooth device power classes*

Power class	Maximum output power	Minimum output power
1	100 mW (20 dBm)	1 mW (0 dBm)
2	2.5 mW (4 dBm)	0.25 mW (−6 dBm)
3	1 mW (0 dBm)	n/a

can be accurately estimated within 0.1 to 0.3 s. These values are for a system suffering a heavy interference level 20 dB above the noise floor. In the case of Bluetooth, the value of SIR depends on the main signal and the interference spectral shape (i.e. whether the main signal falls inside or outside of the interfering signal band). Therefore, given seventy-nine frequency channels, U can be chosen proportionally to four or five times seventy-nine. There is a trade-off between the value of U and the amount of signalling traffic required. A small value for U perhaps allows the system to be more responsive at the cost of having to exchange additional signalling information. In the following example, SIR is measured over an update interval, U, equal to 300 packets, which is equivalent to 0.1875 s.

The power update rule given by Equation (8.4) is implemented at the Bluetooth master and slave devices. Initially, the power is set to $P_{max} = 100$ mW, then updated according to the rule; $P_{min} = 1$ mW. In order to evaluate the performance of this PC algorithm in the presence of interference, let us look at a practical scenario consisting of a four-node topology with two Bluetooth devices and two WLAN (IEEE 802.11b) devices. The Bluetooth devices are placed on a two-dimensional grid at (0, 0) and (1, 0) m, while the the WLAN AP and station are placed at (0, 15) and (0, d) m, respectively. The offered load for the WLAN and Bluetooth connections are set at 50% and 30% of each medium's capacity, respectively. Note that the WLAN transmitted power is fixed at 25 mW. The metrics used to evaluate the performance of the PC algorithm include the probability of packet loss and the mean access delay at the Bluetooth slave device and WLAN station.

First, let us look at how the transmitted power fluctuates at the Bluetooth master device in response to the interference it is subjected to from the WLAN station. Figure 8.2 shows the transmitted power (after $5U$) for the Bluetooth master device versus the distance between the Bluetooth slave from the WLAN station (source of interference). As expected the transmitted power in Figure 8.2 varies between P_{max} and P_{min}. Note that if there is no change in the interference signal, the transmitted power should converge to its final value in one step, i.e. $1U$.

Now let us study the effects of power control on the performance of the Bluetooth system. Figures 8.3(a) and (b) give the packet loss and the access delay,

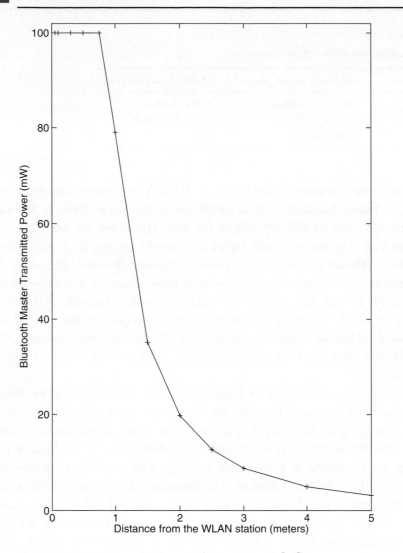

Fig. 8.2. Bluetooth transmitted power (taken from ref. [43]).

respectively, with and without the power control algorithm. For distances equal to 0.5 m from the interference source, increasing the transmitted power leads to lowering the packet loss to ≈4%, instead of 18% without power control. A similar reasoning applies to the delays shown in Figure 8.3(b). However, for distances less than 0.5 m, the transmitted power is capped by $P_{\max}$ and the packet loss remains ≈9%.

A couple of observations are in order. We note that the power control algorithm can be effective in some scenarios; in the case studied here, lower packet losses and access delays are obtained for distances greater than 0.5 m from the interference source. However, it should be made clear that this performance gain comes at the cost of increasing the interference level for other systems. As expected, increasing the Bluetooth transmitted power has a negative impact on the interfering system; in Figure 8.4 we note a 17% packet loss at the WLAN AP device, even if it is about 15 m away from the Bluetooth devices. As the Bluetooth transmitted power is weakened, the packet loss at the WLAN AP device drops to zero.

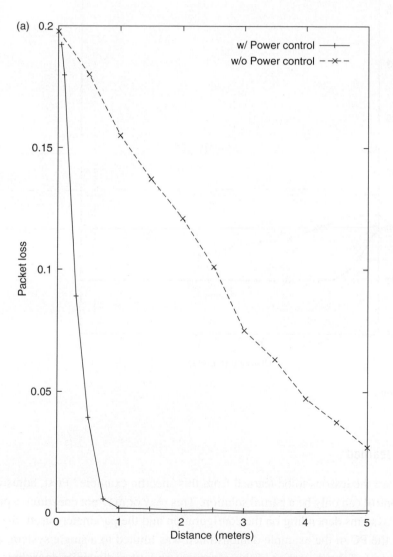

Fig. 8.3. Effects of adaptive power control on Bluetooth performance. (a) Packet loss plotted against distance from the interference source. (b) Mean access delay plotted against distance from the interference source. (Taken from ref. [43].)

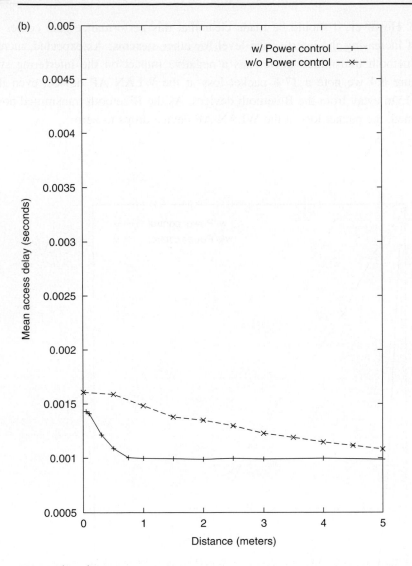

(b)

Fig. 8.3. (cont.).

8.1.2   Lessons learned

So what are the lessons to be learned from this specific example? First, adjusting the power control can only be a partial solution. This may or may not constitute a problem for other systems depending on the configuration and the parameters used. Secondly, although the PC in the example demonstrated was limited to a single system, a more extensive test of the dynamics of this algorithm and other dynamic algorithms for all systems involved does not lead to better results. Thus, PC has limited benefits in the context of interference mitigation.

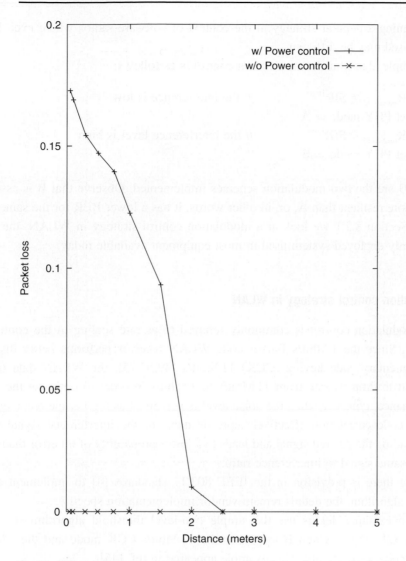

Fig. 8.4. Effects of power control on the other (interference) system. Packet loss is measured on the WLAN AP device (taken from ref. [43]).

8.2   Modulation control

Another common technique widely available in currently deployed systems today is to adjust the modulation scheme according to the operating environment. This is useful, especially in the case where multiple PHY layers with various degrees of robustness are available, in as much as they can all make use of the same MAC layer, as in the case of IEEE 802.11, where there are several PHY schemes available. This is becoming a popular strategy in the context of software radios where even MAC layers used may be different.

A simple algorithm for modulation control is as follows:

1: If $SIR_{measured} \geq SIR^{High}$ // the interference is low
2:      set PHY mode = A
3: If $SIR_{measured} < SIR^{Low}$ // the interference level is high
4:      set PHY mode = B

A and B are the two modulation schemes implemented. Observe that B is assumed to be more resilient than A, or, in other words, it has a lower BER for the same SIR. In the Section 8.2.1 we look at a modulation control strategy in WLAN, the most commonly deployed system used in most equipment available today.

8.2.1   Modulation control strategy in WLAN

This modulation control is commonly referred to as rate scaling in the context of WLAN. Since the 1 Mbit/s Barker code WLAN receiver performs better than the complementary code keying (CCK) 11 Mbit/s [49,63,75], the WLAN data rate is designed to drop its rate from 11 Mbit/s to 1 Mbit/s in order to optimize the range performance, typically when the noise level is perceived as high at the receiver. The Barker code correlation effectively spreads noise or the interference signal while de-spreading the desired signal and leads to a lower probability of bit error than CCK for the same signal to interference ratio.

While there is provision in the IEEE 802.11 standards [4] to implement a rate scaling algorithm, the details remain vendor implementation specific.

As an example, let us use the simple two-level threshold algorithm shown in Section 8.2, where A and B are set to the 11 Mbit/s CCK mode and the 1 Mbit/s Barker code, respectively. This example appeared in ref. [45]:

1: If $SIR_{measured} \geq 6\,dB$ // the interference is low
2:      PHY mode = 11 Mbit/s
3: If $SIR_{measured} < 2\,dB$ // the interference level is high
4:      PHY mode = 1 Mbit/s

The assumption is that when the SIR is low, the interference level is high (or the desired signal is weak), and therefore the receiver reverts to the 1 Mbit/s mode. We set SIR^{High} and SIR^{Low} to 6 and 2 dB, respectively, based on a typical BER performance of each receiver [63]. This so-called hysteresis margin should enable the equipment to avoid unnecessary oscillations related to responding too quickly to instantaneous measurements.

Table 8.2. *Experiment summary*

Experiment	WLAN	Bluetooth
1	FTP	FTP
2	FTP	HTTP

Table 8.3. *Application profile parameters*

Parameters	Distribution	Value
FTP		
Traffic directionality	bi-directional	combination of put and get commands
Inter-request time (s)	exponential	1
File size (bytes)	constant	2M
HTTP		
Page interarrival time (s)	exponential	10
Number of objects per page	constant	2
First object size (bytes)	constant	10 000
Second object size (bytes)	uniform	(2000, 100 000)

Now let us look at the effects of rate scaling on the performance of the WLAN system. We consider a four-node topology consisting of a pair of WLAN devices and a pair of Bluetooth devices. The placement of these nodes on a two-dimensional grid is as follows. The Bluetooth master and slave are placed 1 m apart at $(-0.5, 0)$ and $(0.5, 0)$ m, respectively. The WLAN station is located at $(0, 15)$ m, while the WLAN server is located at $(0, d)$ m; d varies along the y-axis between 0 and 10 m.

Simulations for two different experiments are run and results are collected. For each experiment, the application profiles used for the Bluetooth and WLAN systems are varied as shown in Table 8.2. In experiment 1, both WLAN and Bluetooth use the FTP profile, while in experiment 2 the WLAN application uses FTP while the Bluetooth application uses HTTP. The FTP and HTTP parameters used are described in Table 8.3.

We compare the performance of WLAN and Bluetooth when rate scaling is used for WLAN. Two sets of simulations are conducted in order to identify the benefits of the rate scaling algorithm. In Figure 8.5, "None" refers to the case when no algorithm is used, and "Rate scaling" means that WLAN uses the rate scaling algorithm.

First, let us study the performance of the WLAN system. Figure 8.5(a) gives the packet loss with respect to the y-coordinate of the WLAN server, d, when both WLAN and Bluetooth use the FTP profile. When no algorithm is used, the packet loss can be up to 14% when the WLAN server is close to the Bluetooth piconet

($d = 0$ m). As the server moves away from the Bluetooth piconet, the packet loss drops to zero (d ≥ 5 m). With rate scaling, the packet loss starts at 5% when $d = 0$ m. This observed packet loss is due to the intermittent use of the 11 Mbit/s WLAN receiver before the 1 Mbit/s mode is used. While the adaptive filter used in the 1 Mbit/s receiver is able to reduce the packet loss to zero, the 11 Mbit/s receiver is less robust and yields a relatively high packet loss.

Figure 8.5(b) illustrates the TCP throughput of the WLAN server. When no algorithm is used, the throughput starts at 240 kbyte/s when $d = 0$ m, and increases

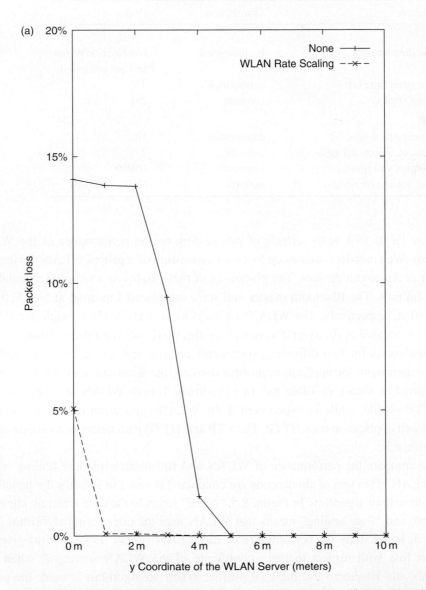

Fig. 8.5. Experiment 1: WLAN FTP performance. (a) Packet loss. (b) TCP throughput.

(b)

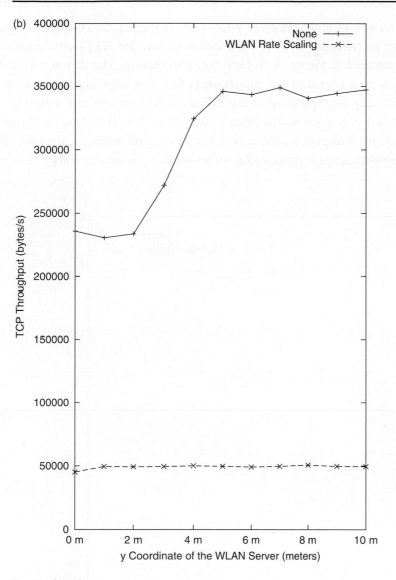

Fig. 8.5. (cont.).

to 350 kbyte/s when $d \geq 5$ m and the packet loss is zero. Observe that when BIAS is used, the throughput remains around 350 kbyte/s since no packets are lost. Since rate scaling involves reducing the WLAN bit rate from 11 to 1 Mbit/s, this amounts to reducing the throughput to 50 kbyte/s. As expected, rate scaling can reduce the packet loss, at the cost of reducing the throughput.

Overall, we note that the use of rate scaling produces interesting but expected trade-offs. While the WLAN packet loss is reduced, the delay is increased and the throughput is reduced.

Next, let us look at the effects of WLAN rate scaling on the Bluetooth system. The results for packet loss and delay when Bluetooth uses the HTTP profile (experiment 2) are illustrated in Figures 8.6(a) and (b), respectively. The reason why the delay measure is used instead of the throughput is that it is more important to the HTTP application than the throughput measure. The packet loss with rate scaling is slightly higher (11%) than when no algorithm is used (8%). The TCP delay in Figure 8.6(b) starts at 33 ms with rate scaling at $d = 0$ m; it is 12 ms when no algorithm is used. When no interference is present ($d = 10$ m) the delay is around 6 ms.

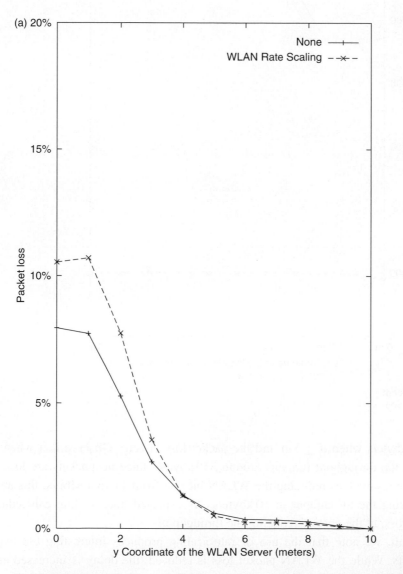

Fig. 8.6. Experiment 2: Bluetooth HTTP performance. (a) Probability of packet loss. (b) TCP delay.

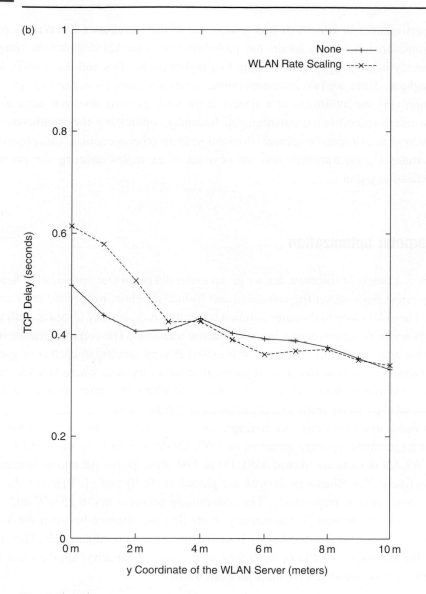

Fig. 8.6. (cont.).

Clearly the use of rate scaling for WLAN leads to higher packet losses for Bluetooth, including higher delays and a lower throughput.

8.2.2   Lessons learned

Let us summarize the lessons learned from this example. The benefits of using rate scaling in the WLAN system lead to mixed results. While the packet loss is reduced for WLAN due to the use of a more robust receiver and an adaptive filter, the performance of Bluetooth is degraded due to the increase of the WLAN packet transmission time. As a result, the probability of a packet collision in time and frequency is more significant, leading to a higher packet loss and delay and a lower throughput. There are two important points to observe here. First, the strategy aimed at improving the robustness of a system in the case of noise does not work as well in an interference limited environment. Secondly, optimizing the parameters for a given system in a specific scenario does not scale to other scenarios. This observation constitutes a good transition into the next set of examples covering the parameter optimization pitfall.

8.3   Parameter optimization

There is a family of solutions that we group under the parameter optimization heading. They range from packet fragmentation and format selection, to forward error correction. They all belong to the same solution space that is essentially aimed at optimizing the choice of some parameters in some specific scenarios. The common feature in this solution space is that optimization, if it exists, is very specific to each configuration and therefore rarely scales well. It seems that for every case where benefits can be clearly identified, there is at least one other case where the same optimal set has no effect, or even worse leads to a degradation in the performance.

In order to demonstrate this concept, we use a series of simulation scenarios all using a four-node topology including two WLAN devices and two Bluetooth devices. The WLAN devices are located at $(0, 15)$ and $(0, d)$ m for the AP and mobile device, respectively. The Bluetooth devices are placed at $(0, 0)$ and $(1, 0)$ m for the slave and master device, respectively. The transmitting power is set to 25 mW and 1 mW for WLAN and Bluetooth, respectively. Note that the distance between the WLAN mobile node and the Bluetooth slave is varied along the y-coordinate axis. The offered load for the WLAN traffic is set to 50% of its channel capacity. On the other hand, a 30% load is considered for the Bluetooth traffic.

8.3.1   Effects of packet fragmentation

It is generally accepted that packet fragmentation leads to better results for the interferer system. This is simply due to the fact that smaller packets have a lower probability of packet collision. Intuitively, since smaller packets spend less time "in the air", they are less prone to collisions. For an in-depth analysis of the relationship between the packet size, the packet interarrival time, and the probability of packet error, the reader is referred to Chapter 4.

Now let us look at the effects of packet fragmentation on the victim system. As we expected, fragmentation leads to more collisions on the victim signal. Figure 8.7

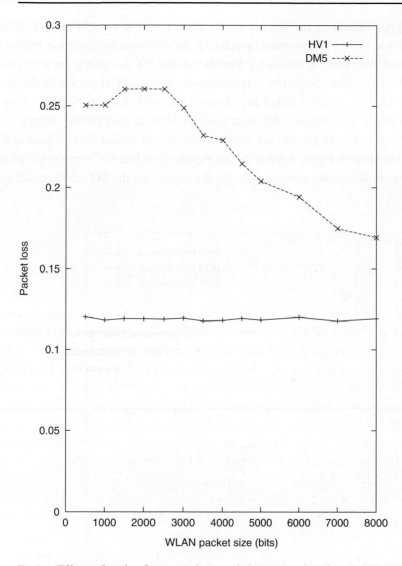

Fig. 8.7. Effects of packet fragmentation on victim system: interferer is WLAN; victim is Bluetooth.

shows the effects of using different WLAN packet sizes on the Bluetooth packet loss. While varying the WLAN packet size has no effect on the Bluetooth short HV1 packet types, there is a significant difference on the Bluetooth DM5 type packets. DM5 packets that are five times longer than HV1 packets are twice as much affected by the shorter WLAN packet size.

Following a similar argument, let us look at the effects of varying the interval between two Bluetooth packet transmissions on the WLAN packet loss. Three different types of Bluetooth voice packets are used, namely HV1, HV2, and HV3 which are sent every slot, every two slots, and every four slots. Observe that HV packet types belong to the synchronous connection oriented (SCO) link definition in Bluetooth. SCO is a symmetric point-to-point connection between a master and a slave where the master sends an SCO packet in one TX slot at regular time intervals, defined by T_{SCO} time slots. The slave responds with an SCO packet in the next TX opportunity. T_{SCO} is set to either two, four or six time slots for HV1, HV2, or HV3 packet formats, respectively. All three formats of SCO packets are defined to carry 64 kbits/s of voice traffic and are never retransmitted in the case of packet loss or error. Observe from Figure 8.8 that the smaller the time interval between the Bluetooth packet transmission, the more significant the impact on the WLAN's packet loss.

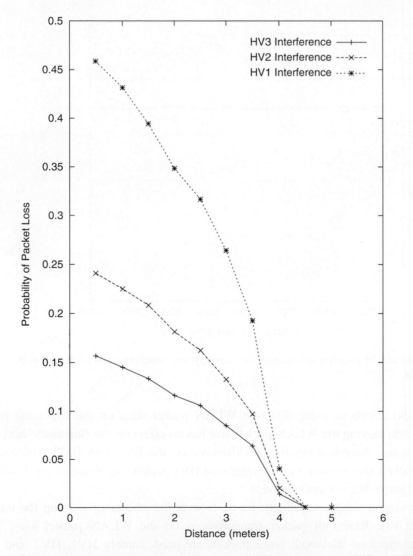

Fig. 8.8. Effects of packet transmission interval on victim system: interferer is Bluetooth; victim is WLAN.

8.3.2   Effects of forward error correction

Now let us investigate the benefits of error correction for the victim system. By repeating the same experiments using different packet encapsulation types, namely DM and DH for Bluetooth, we can isolate the effects of FEC on the Bluetooth performance. The experiments are repeated for three different packet sizes, DM1/DH1, DH3/DM3, and DH5/DM5, which occupy one, three, and five slots, respectively. These packet formats are described in greater detail in Chapter 2. The offered load for Bluetooth is set to 30% of the channel capacity, which corresponds to a packet interarrival of 2.91 ms, 8.75 ms and 14.58 ms for DH1/DM1, DH3/DM3, and DH5/DM5 packets, respectively. In this case, we note from Figure 8.9 that the use of FEC has limited benefits and can only

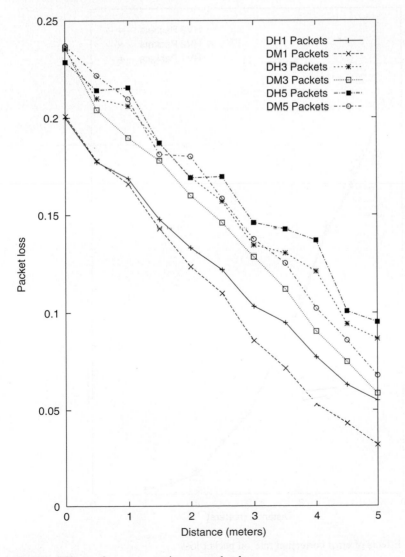

Fig. 8.9. Effects of error correction on packet loss.

improve the performance of Bluetooth for low interference scenarios (i.e. for distances when the WLAN device is more than 3 m away).

We further explore the effects of the error correction rate. We vary the error correction rate from HV1, which uses a 1/3 FEC rate and a $T_{SCO} = 2$, to HV2, which uses a 2/3 FEC rate and a $T_{SCO} = 4$, and HV3, which uses no FEC and a $T_{SCO} = 6$. Note that there is no difference in the total packet length between the different HV packets. From Figure 8.10, we observe that the choice of packet encapsulation does not impact the performance of Bluetooth; in other words, the use of additional error correction does not improve performance. On the other hand, note that in the previous result from Figure 8.8 the HV3 is "friendlier" to WLAN due to a longer T_{SCO} period.

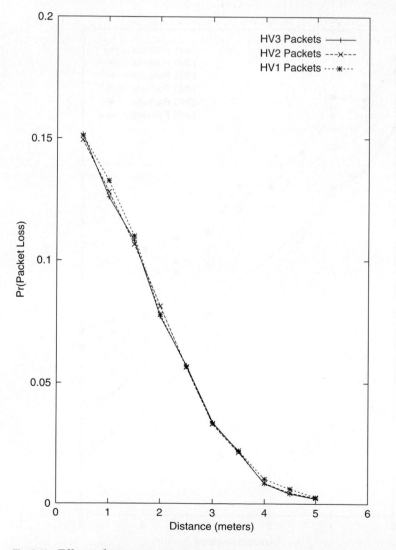

Fig. 8.10. Effects of error correction rate on packet loss

8.3.3   Lessons learned

We give several examples regarding the use of FEC, the choice of packet encapsulation and fragmentation, and their effects on performance. Our observations include the following. First, the use of FEC has limited benefits for many interfering scenarios. Secondly, applying fragmentation can reduce the probability of packet loss at the expense of causing more interference on the "other" system. Beyond these observations, it is useful to point out that tuning parameters to achieve acceptable performance for a particular system often comes at the expense of the other system's throughput. Moreover, there is little room for parameter optimization, especially for practical scenarios. Not only does the complexity of the interactions and the number of parameters that have to be adjusted make the optimization problem intractable, but choosing an objective function is very dependent on the applications and the scenario. Therefore, parameter optimizations should not constitute a primary focus in interference mitigation solutions.

References

Standard specifications

[1] Bluetooth Special Interest Group. Specifications of the Bluetooth system, vol. 1, v.1.0B 'Core' and vol. 2, v.1.0B 'Profiles' (1999).

[2] Federal Communications Commission, Title 47, Code for federal regulations, part 15 (1998).

[3] HomeRF Working Group, HomeRF shared wireless access protocol cordless access (SWAP-CA) specifications (2000).

[4] IEEE Std 802.11-1999/8802-11 (ISO/IEC 8802-11:1999). IEEE Standard for information technology – LAN/MAN – Specific requirements – Part 11: Wireless LAN medium access control (MAC) and physical layer (PHY) specifications (1999).

[5] IEEE Std 802.11A-1999/8802-11, A (ISO/IEC 8802-11/Amd 1). Information technology – Telecommunications and information exchange between systems – LAN/MAN – specific requirements (1999).

[6] IEEE Std 802.11B-1999. IEEE local and metropolitan area networks – Specific requirements – Part 11: Wireless LAN medium access control (MAC) and physical layer (PHY) specifications: Higher speed physical layer (PHY) extension in the 2.4 GHz band (1999).

[7] IEEE Std 802.15.1-2002. IEEE Standard for information technology-telecommunications and information exchange between systems – Local and metropolitan area networks – Specific requirements (2002).

[8] IEEE Std 802.11G-2003. IEEE Standard for IT-telecommunications and information exchange between systems LAN/MAN – Part II: Wireless LAN medium access control (MAC) and physical layer (PHY) specifications amendment 4: Further higher data rate extension in the 2.4 GHz band (2003).

[9] IEEE Std 802.15.2-2003. IEEE recommended practice for telecommunications and information exchange between systems – Local and metropolitan area networks specific requirements – Part 15.2: Coexistence of wireless personal area networks with other wireless devices operating in unlicensed frequency band (2003).

[10] IEEE Std 802.15.3-2003. IEEE Standard for information technology – Telecommunications and information exchange between systems – Local and metropolitan area networks – Specific requirements – Part 15.3: Wireless medium access control (MAC) and physical layer (PHY) specifications for high rate wireless personal area networks (WPAN) (2003).

[11] IEEE Std 802.15.4-2003. IEEE Standard for information technology – Telecommunications and information exchange between systems – Local and metropolitan area networks – Specific requirements – Part 15.4: Wireless medium access control (MAC) and physical

layer (PHY) specifications for low rate wireless personal area networks (LR-WPANs) (2003).

[12] IEEE Std 802.1X-2004. IEEE Standard for local and metropolitan area networks port-based network access control (2004).

[13] IEEE Std 802.11I-2004. IEEE Standard for information technologyTelecommunications and information exchange between systems – LAN/MAN – Specific requirements – Part 11: Wireless LAN medium access control (MAC) and physical layer (PHY) specifications amendment 6: MAC security enhancements (2004).

[14] IEEE Std 802.16-2004. IEEE Standard for local and metropolitan area networks – Part 16: Air interface for fixed broadband wireless access systems (2004).

[15] Infrared Data Association, IrDA Advanced infrared physical layer specification, v.1.0 (1998).

[16] Joint Technical Committee T1 R1P1.4 and TIA TR46.3.3/TR45.4.4 on Wireless access, draft final report on RF channel characterization, Paper no. JTC(AIR)/94.01.17-238R4 (1994).

[17] Wireless Communications Systems – Performance in noise and interference limited situations – recommended methods for technology independent modeling, simulation and verification, *TIA/EIA Telecommunications Systems Bulletin*, **TSB88-A** (1999).

Standard group contributions

[18] G. Ennis, Impact of Bluetooth on 802.11 direct sequence, *IEEE P802.11 Working Group Contribution*, IEEE P802.11-98/319 (1998).

[19] N. Golmie, Interference aware Bluetooth scheduling techniques, *IEEE P802.11 Working Group Contribution*, IEEE P802.15-01/143r0 (2001).

[20] N. Golmie and F. Mouveaux, Impact of interference on the Bluetooth access control performance: preliminary results, *IEEE P802.15 Working Group Contribution*, IEEE P802.15-00/322r0 (2000).

[21] A. Kamerman, Coexistence between Bluetooth and IEEE 802.11 CCK: solutions to avoid mutual interference, *IEEE P802.11 Working Group Contribution*, IEEE P802.11-00/162r0 (2000).

[22] J. Lansford, R. Nevo and E. Zehavi, MEHTA: a method for coexistence between co-located 802.11b and Bluetooth systems, *IEEE P802.11 Working Group Contribution*, IEEE P802.15-00/360r0 (2000).

[23] S. Shellhammer, Packet error rate of an IEEE 802.11 WLAN in the presence of Bluetooth, *IEEE P802.15 Working Group Contribution*, IEEE P802.15-00/133r0 (2000).

[24] B. Treister, A. Batra, K. C. Chen and O. Eliezer, Adaptative frequency hopping: a non-collaborative coexistence mechanism, *IEEE P802.11 Working Group Contribution*, IEEE P802.15-01/252r0 (2001).

[25] J. Zyren, Extension of Bluetooth and 802.11 direct sequence model, *IEEE P802.11 Working Group Contribution*, IEEE P802.11-98/378 (1998).

[26] J. Zyren, Reliability of IEEE 802.11 WLANs in presence of Bluetooth radios, *IEEE P802.11 Working Group Contribution*, IEEE P802.15-99/073r0 (1999).

Journal and conference articles

[27] M. Andersin, N. Mandayan and R. Yates, A subspace based estimation of the signal to interference ratio for TDMA cellular systems, *Proceedings of IEEE Vehicular Technology Conference*, VTC'96 (1996), pp. 1155–1159.

[28] N. Bambos, Toward power-sensitive network architectures in wireless communications: concepts, issues, and design aspects, *IEEE Personal Communications*, **5** (1998), 50–59.

[29] P. Bhagwat, P. Bhattacharya, A. Krishna and S. Tripathi, Enhancing throughput over wireless LANs using channel state dependent packet scheduling, *Proceedings of IEEE INFOCOM* (1996), pp. 1133–1140.

[30] C. Bisdikian, A review of random access algorithms, *Proceedings of International Workshop on Mobile Communications* (1996), pp. 123–127.

[31] N. Chevrollier, N. Montavont and N. Golmie, Handovers and interference mitigation in healthcare environment, *Proceedings of IEEE Military Communications Conference (MILCOM)*, MILCOM 2005, Atlantic City, NJ, October 17–20, 2005.

[32] C. F. Chiasserini and R. R. Rao, Coexistence mechanisms for interference mitigation between IEEE 802.11 WLANs and Bluetooth, *Proceedings of IEEE INFOCOM* (2002), pp. 590–598.

[33] D. Chiu and R. Jain, Analysis of the increase and decrease algorithms for congestion avoidance in computer networks, *Computer Networks and ISDN Systems*, **17** (1989), pp. 1–14.

[34] D. Eckhardt and P. Steenkiste, Measurement and analysis of the error characteristics of an in-building wireless network, *Proceedings of ACM Special Interest Group on Communications (SIGCOMM)* (New York: ACM Press, 1996), pp. 243–254.

[35] D. Eckhardt and P. Steenkiste, A trace-based evaluation of adaptive error correction for a wireless local area network, *Mobile Networks and Applications (MONET)*, **4** (1999), 273–287.

[36] S. Floyd and V. Jacobson, Link sharing and resource management models for packet networks, *ACM/IEEE Transactions on Networking*, **3** (1995), 365–386.

[37] G. Foschini and Z. Miljanic, A simple distributed autonomous power control algorithm and its convergence, *IEEE Journal on Vehicular Technology*, **42** (1993), 641–646.

[38] C. Fragouli, V. Sivaraman and M. B. Srivastava, Controlled multimedia wireless link sharing via enhanced class-based queueing with channel-state-dependent packet scheduling, *Proceedings of IEEE INFOCOM* (1998), pp. 572–580.

[39] B. D. Fritchman, A binary channel characterization using partitioned Markov chains, *IEEE Transactions on Information Theory*, **13** (1967), 221–227.

[40] D. Fumolari, Link performance of an embedded Bluetooth personal area network, *Proceedings of IEEE International Conference on Communications*, ICC 2001, pp. 2573–2577.

[41] E. N. Gilbert, Capacity of a burst-noise channel, *Bell Systems Technical Journal*, **39** (1960), 1253–1266.

[42] N. Golmie, Bluetooth dynamic scheduling and interference mitigation, *ACM Mobile Networks and Applications (MONET)*, **9** (2004), 21–31.

[43] N. Golmie and N. Chevrollier, Techniques to improve Bluetooth performance in interference environment, *Proceedings of IEEE Military Communications*, MILCOM 2001, pp. 581–585.

[44] N. Golmie and F. Mouveaux, Interference in the 2.4 GHz ISM band: impact on the Bluetooth access control performance, *Proceedings of IEEE International Conference on Communications*, ICC 2001, pp. 2540–2545.

[45] N. Golmie and O. Rebala, Techniques to improve the performance of TCP in a mixed Bluetooth and WLAN environment, *Proceedings of IEEE International Conference on Communications*, ICC 2003, pp. 1181–1185.

[46] N. Golmie, N. Chevrollier and O. Rebala, Bluetooth adaptive frequency hopping and scheduling, *Proceedings of Military Communications*, MILCOM 2003, pp. 405–409.

[47] N. Golmie, N. Chevrollier and O. Rebala, Bluetooth and WLAN coexistence: challenges and solutions, *IEEE Wireless Communication Magazine*, **10** (2003), 22–29.

[48] N. Golmie, R. E. Van Dyck and A. Soltanian, Interference of Bluetooth and IEEE 802.11: simulation modeling and performance evaluation, *Proceedings of the Fourth ACM International Workshop on Modeling, Analysis, and Simulation of Wireless and Mobile Systems (MSWIM)* (New York: ACM Press, 2001), pp. 11–18.

[49] N. Golmie, R. E. Van Dyck, A. Soltanian, A. Tonnerre and O. Rebala, Interference evaluation of Bluetooth and IEEE 802.11b systems, *ACM Wireless Networks (WINET)*, **9** (2003), 202–211.

[50] H. Hashemi, The indoor radio propagation channel, *Proceedings of the IEEE*, **81** (1993), 943–968.

[51] I. Howitt, WLAN and WPAN coexistence in UL band, *IEEE Transactions on Vehicular Technology*, **50** (2001), 1114–1124.

[52] I. Howitt, V. Mitter and J. Gutierrez, Empirical study for IEEE 802.11 and Bluetooth interoperability, *Proceedings of IEEE Vehicular Technology Conference*, VTC'01 2001, pp. 1109–1113.

[53] D. Johnson, C. Perkins and J. Arkko, Mobility support in IPv6, *Internet Engineering Task Force Request for Comments (RFC) 3775*; www.ietf.org.

[54] Y. Y. Kim and S. Q. Li, Modeling multipath fading channel dynamics for packet data analysis, *Proceedings of IEEE Infocom*, (1998), 1292–1300.

[55] J. Lansford, S. Stephens and R. Nevo, Wi-Fi (802.11b) and Bluetooth: enabling coexistence, *IEEE Network Magazine*, **15** (2001), 20–27.

[56] J. D. Laster and R. H. Reed, Interference rejection in digital wireless communications, *IEEE Signal Processing Magazine*, **14** (1997), 37–62.

[57] L. Lu, V. Bharghawan and R. Srikant, Fair scheduling in wireless packet networks, *Proceedings of ACM Special Interest Group on Communications (SIGCOMM)* (New York: ACM Press, 1997), pp. 63–74.

[58] T. S. E. Ng, I. Stoica and H. Zhang, Packet fair queueing algorithms for wireless networks with location dependent errors, *Proceedings of INFOCOM*, (1998), pp. 1103–1111.

[59] G. T. Nguyen, A trace-based approach to evaluating wireless networks, M.S. thesis, University of California at Berkeley (1996).

[60] B. Noble, M. Satyanarayanan, G. T. Nguyen and R. H. Katz, Trace-based mobile network emulation, *Proceedings of ACM Special Interest Group on Communications (SIGCOMM)* (New York: ACM Press, 1997), pp. 51–61.

[61] J. del Prado and S. Choi, Experimental study on coexistence of 802.11b with alien devices, *Proceedings of IEEE Vehicular Technology Conference*, VTC'01 (2001), pp. 977–981.

[62] P. Ramanathan and P. Agrawal, Adapting packet fair algorithms to wireless networks, *Proceedings of ACM/IEEE Mobile Communications (MOBICOM)* (New York: ACM Press, 1998), pp. 1–9.

[63] A. Soltanian and R. E. Van Dyck, Physical layer performance for coexistence of Bluetooth and IEEE 802.11b, *Proceedings of Virginia Tech Symposium on Wireless Personal Communications*, Blacksburg, Va, June 6–8, 2001.

[64] F. A. Tobagi and L. Kleinrock, Packet switching in radio channels. Part ii: The hidden terminal problem in carrier sense multiple-access and the busy-tone dial solution, *IEEE Transactions on Communications*, **23** (1975), 1417–1433.

[65] H. S. Wang and N. Moayeri, Finite-state Markov channel – a useful model for radio commu- nication channels, *IEEE Transactions on Vehicular Technology*, **44** (1995), 163–171.

[66] J. Zander, Distributed cochannel interference control in cellular radio systems, *IEEE Trans- actions on Vehicular Technology*, **41** (1992), 305–311.

[67] S. Zurbes, W. Stahl, K. Matheus and J. Haartsen, Radio network performance of Bluetooth, *Proceedings of IEEE International Conference on Communications*, ICC 2000, pp. 1563–1567.

Books

[68] H. L. Bertoni, *Radio Propagation for Modern Wireless Systems* (Englewood Cliffs, NJ: Prentice Hall, 2000).

[69] D. Bertsekas and R. Gallager, *Data Networks, 2nd edn* (Prentice Hall, 1992).

[70] S. Haykin, *Communication Systems, 4th edn* (New York: Wiley, 2001).

[71] W. C. Jakes, *Microwave Mobile Communications* (New York: Wiley Interscience, 1974).

[72] M. C. Jeruchim, P. Balaban and K. C. Shanmugan, *Simulation of Communication Systems, Modeling, Methodology and Techniques, 2nd edn*, (New York: Kluwer Academic/Plenum Publisher, 2000).

[73] T. S. Rappaport, *Wireless Communications: Principles and Practice* (Englewood Cliffs, NJ: Prentice Hall, 1996).

[74] M. Schwartz, *Telecommunication Networks: Protocols, Modeling and Analysis* (Reading, MA: Addison-Wesley, 1988).

[75] B. Sklar, *Digital Communications: Fundamentals and Applications, 2nd edn* (Englewood Cliffs, NJ: Prentice Hall, 2001).

[76] A. Viterbi, *CDMA: Principles of Spread Spectrum Communication* (Englewood Cliffs, NJ: Prentice Hall PTR, 2005).

[77] B. H. Walke, *Mobile Radio Networks: Networking and Protocols* (New York: John Wiley and Sons, 1999).

Index